Taking Things Apart & Putting Things Together

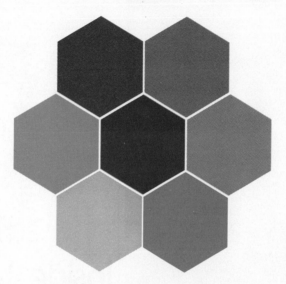

by John H. Woodburn

Design and graphics
Joe Phillips
Base Guiley

Published by the American Chemical Society
in recognition of its Centennial 1876-1976

American Chemical Society
1155 16th Street N.W., Washington, D.C. 20036

Contents

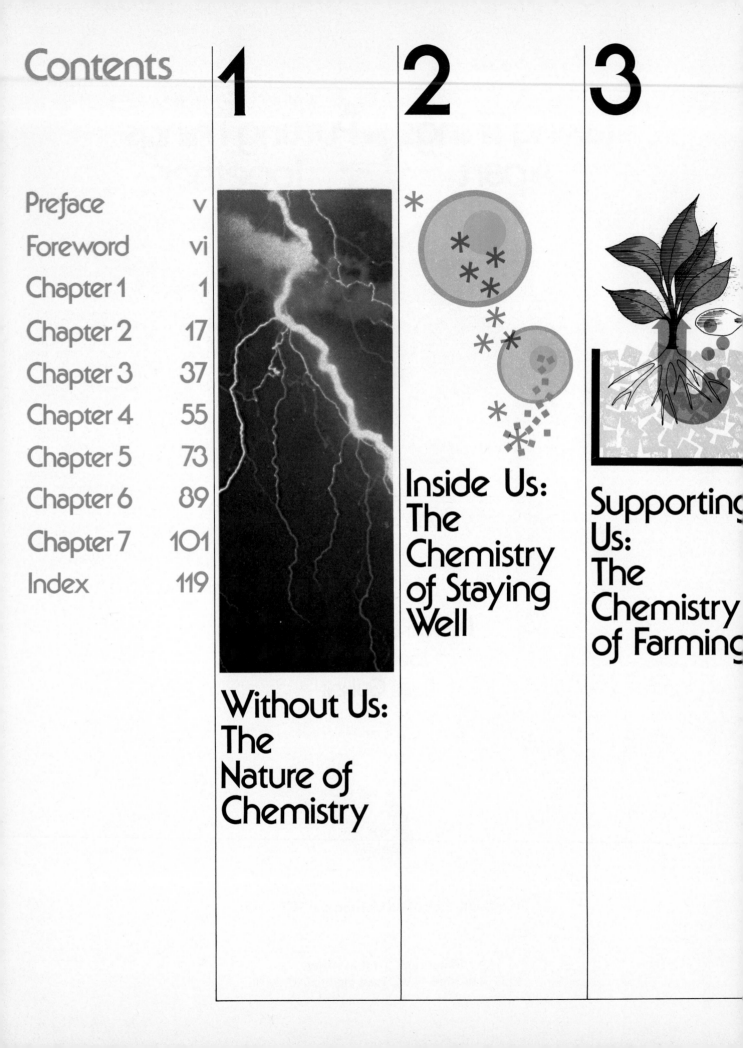

1

Without Us:
The
Nature of
Chemistry

2

Inside Us:
The
Chemistry
of Staying
Well

3

Supporting
Us:
The
Chemistry
of Farming

4

Around Us: What Chemists Make for People

5

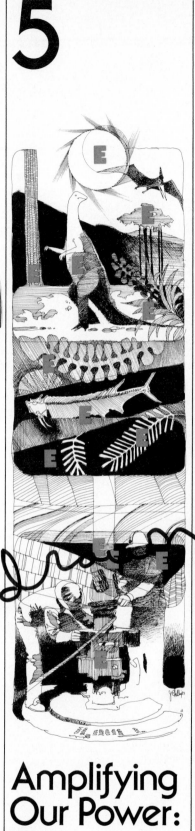

Amplifying Our Power: The Chemistry of Energy

6

Extending Our Senses: Seeing Further With Chemistry

7

Choices Through Chemistry

Library of Congress Cataloging in
Publication Data

Woodburn, John H
  Taking things apart and putting
  things together.

  Includes index.

  1. Chemistry—Popular works.
  I. Title.

QD37.W63   540   76-20448
ISBN 0-8412-0314-8

Preface

This is the story of what chemistry is, what chemists do, and what the results have been.

The title and much of the spirit of the book follow the American Chemical Society's Centennial Exhibit that was designed and produced by Chermayeff & Geismar Associates, with the text by Ralph Caplan.

We know, of course, that the story of chemistry is far too vast to tell in a single book, but we do want to provide glimpses of a fascinating field of knowledge and know-how that is involved intimately in your well-being, your way of making a living, your universe.

Each episode in chemistry is a series of events stemming from a magnificently conceived set of building blocks, the atoms of the earth's elements. These building blocks can come together in supremely orderly fashion to yield millions of substances with properties so varied as to fulfill just about any need imaginable. Most miraculous of all are those substances which not only reproduce themselves but are organized into the fantastically complex systems of plants and animals, including people.

We know that chemistry is an inseparable but not yet perfectly coordinated part of our total society. Forces of change have been unleashed sometimes too soon; at other times, too late. However, the chemical profession and industry try always to be mindful of the consequences.

Looking back even before the birth of the Society, we see that people change their feelings about chemistry. Public attitudes have grown from ignorant suspicion to great expectations and appreciation, and have now become somewhat doubtful again. While his students were helping to develop this book, the author was reminded constantly of these changing attitudes. He is especially indebted to one young man who seemed to be schooled well beyond his years in the ways of the political world. With only enough tongue-in-cheek to make sure his comment would not be misinterpreted, he cautioned that this book might become "deceptively simple, bordering on the naive scheme of propaganda designed to enslave the people to the monopolistic octopus of the Chemical Industry."

Chemistry is the science and technology of taking things apart and putting things together. This definition could be rephrased to say that chemistry is reshuffling atoms from molecule to molecule—but this version takes us into the domain of the unseeable. Quoting another student, "Chemistry is a science of assumptions. We have no way of seeing what we are talking about. We can work only from what we say chemistry is. We propose hypotheses and theories but always against the framework of what we assume chemistry to be. When the things we *can* see seem to be true manifestations of the interactions between atoms and energy we anticipated, then we decide our assumptions make good sense."

It is both an advantage and a disadvantage for this to be a 100th birthday book. Some of the discoveries, the inventions, the products and processes of chemistry need the passage of time before their stories can be told. But "old hats" are seldom as exciting as new. In any case, the book brings together episodes from the past as well as the present. We even indulge now and then in looking to the future.

Perhaps some of the episodes will illuminate motivations of chemists, as well as the way chemists approach problems, the kinds of evidence they believe to be significant, and how they arrive at choices and decisions. It would be foolish to claim that we are revealing formulas for solving problems. It may be, however, that what it takes to figure out and manage the behavior of atoms can also help shape up and solve other kinds of problems.

Foreword

This is a special year for all Americans, as we celebrate in a variety of ways, the Bicentennial of the American Revolution. For the nation's chemists, however, 1976 has added significance because it marks the Centennial of the American Chemical Society. The coincidence of the two anniversaries serves to emphasize the fact that the nation owes much of its progress to the contributions of chemistry over the past 100 years.

To help the public understand the nature and extent of these contributions is the primary purpose of this book and of the exhibit on which it is based.

As a chemist who often has had the opportunity to interpret chemistry to the public, especially when I was chairman of the Atomic Energy Commission, I have been impressed by the public's sincere desire to learn more about science and its effects on everyday life.

To many people, chemistry seems to be a difficult subject and the work of chemists hard to understand. Much of this may be due to the language of chemistry. Many of my lectures that have been most satisfying to me have been those in which I have talked to people about chemistry without using the technical language of scientific seminars.

Dr. John H. Woodburn, author of this book, has spent years writing about science for the general public. We are fortunate to have persuaded him to write a book to enhance the usefulness of our Centennial Exhibit, entitled "Taking Things Apart and Putting Things Together." This story, dramatizing what chemistry is, what chemists do, and what the results have been, is told lucidly in both the exhibit and the book.

You may have noticed that the word chemistry does not appear in the title of this book. Actually, the word could well appear on every page of this and many other books. Chemistry is a pervasive enterprise. More important than to emphasize the word is to urge that you read this book; thereafter chemistry will have a special niche in your memory system.

Glenn T. Seaborg
President
American Chemical Society, 1976

1
Without Us: The Nature of Chemistry

Changing, changing, our world is always changing. In nature, old things are constantly being exchanged for new. Plants change the substances they absorb from the air and soil into starches and cellulose, sugars and fats, proteins and vitamins and everything else we find in roots, stems, leaves, flowers, seeds, and fruits. Animals take apart the substances in their food and put together new bone and muscle, nerves and fat, hair and horn, milk and skin.

Nature's chemistry is never turned off. The bunsen burner is always being adjusted. Gravitational forces tumble anything that can fall or flow. Winds and waves are always moving things about, mixing and stirring nature's reactants as vigorously as any student ever shakes a test tube. Water repeats endlessly its cycle between clouds and streams, lakes and oceans. Slowly trickling groundwater extracts soluble minerals from exposed rocks to become inexhaustible solutions. Streams of different solutions join and new substances are created.

Crashing streaks of lightning accompany massive changes in the electrical balance of materials in nature—changes likely to result in objects being taken apart explosively. The more gentle sparks of static electricity that jump between things rubbed together bear further witness to the electrical nature of materials and substances.

Nature's chemistry is fantastically successful. Pigments occur in all colors of the rainbow. There are countless fragrances and flavors, both pleasant and unpleasant. There are substances as hard as diamond or as soft as wool, as black as coal or as clear as ice, as life-saving as oxygen or as deadly as cobra venom. In short, nature's world includes an almost infinite variety of materials.

The chemistry of nature is fascinating. Snowflakes and mineral crystals are models of symmetry and orderliness. The "recipes" whereby nature's chemicals are put together are measured down to each individual atom or molecule. Chemists know they can "hang their hats on" the uniformity of melting and boiling temperatures of substances, the density of pure samples of materials, and many other characteristics of substances.

But people are not content simply to admire nature's chemistry. They want to understand. People are uneasy when they are dependent on actions they don't understand and, consequently, can neither manage nor control. People want a piece of the action; they want to be free from blind dependence on nature's chemistry. They want to progress, to make better use of or even go beyond nature's chemistry.

Our interests in chemistry, in turn, come from our needs and concerns, our aches and urges, drives and desires. Chemistry's story consists of the efforts of people to fulfill their needs, to develop their interests, to cure their aches and pains, and to know the satisfactions of living in a world they understand.

Sometimes it is good to look at things through the framework of 100-year or centennial birthdays. This helps us see how things grow and develop. It may even help us anticipate and prepare for the future. In 1976, the United States celebrates its two-hundredth and the American Chemical Society its one-hundredth birthday. It is interesting to wonder what chemistry was like, what chemists did, what the results of chemistry were before 1776. How did things change between 1776 and 1876? Between 1876 and 1976? And if we like to dream, what will chemistry be like and what will chemists be doing in 2076? Or 20,076?

Questions which involve the future are especially tough, but none of the above questions is easy to answer. For one reason, chemistry is only one of the things people do. It isn't easy to untangle chemistry from all of the other things people do; people such as artists and engineers, musicians and politicians, plumbers and preachers. To really understand what chemistry is, what it has done in the past and may do in the future, we would have to look at chemistry as a part of the total stream of civilization. Perhaps then we could sense the unique role of the men and women who make careers of taking things apart and putting things together.

Rocks and soil are nature's chemicals.
So are the gases of the atmosphere and
all materials dissolved in streams and oceans

We must stretch our imaginations to appreciate the quantities of nature's chemicals. Try these "mind stretchers," for example. There is enough water in the world to fill a 25,000 gallon swimming pool for each of the world's people more than three million times; enough gold dissolved in the world's oceans to build 10 solid gold automobiles for each of the world's four billion people; enough air over a city block to let each person fill his or her lungs with at least one breath of fresh air.

Nature is not so well supplied with some kinds of elements, the basic building blocks of the world. The story is told of a boy who hoped his chemist friends would give him one-ounce samples of all of the earth's 92 elements. The boy was surprised to learn that some of the elements are so costly that a one-ounce sample would cost thousands of dollars. Furthermore, efforts to obtain one-ounce samples of some of the rarer elements would seriously deplete the world's supply.

It was good that the boy was interested in the earth's building blocks, and a collection of the elements would make a valuable and informative collection. Although the boy was disappointed, he learned a lesson more people should learn when we dip into nature's chemicals to fulfill a need or to solve a problem. This is one of the painful lessons being taught us by the impending shortage of petroleum and natural gas. The supplies of nature's materials are finite.

The language of chemistry

Perhaps we should describe how we use some of the words that help explain what chemistry is and what chemists do.

Raw materials are volumes of liquids or gases or chunks of solids which provide the building blocks chemists need to put together the substances they want. The term *building block* is not meant to be a precise term, but

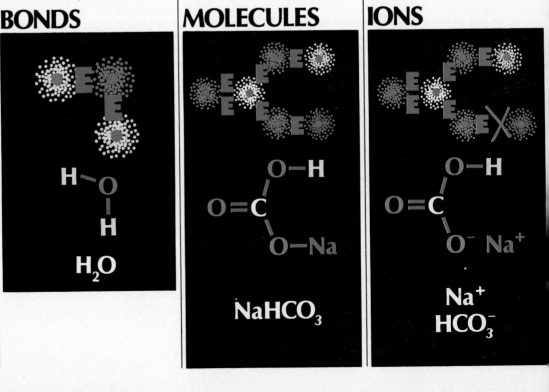

Chemists use symbols to represent atoms, the individual building blocks of elements.

Structural formulas are used to show how atoms are arranged in molecules. In structural formulas, short lines represent bonds between atoms. Each kind of atom has a definite number of electrons it can share with other atoms. On this basis, for example, hydrogen atoms can share only a single bond, oxygen atoms can form two single bonds or one double bond, carbon atoms can form four single, two double bonds, a triple and a single bond, or any combination that is equivalent to four single bonds. In some cases, how many electrons an atom has to share depends upon the kinds of other atoms with which it is bonding.

The individual particles of many kinds of materials break up when they go into solution. If the fragments are electrically charged they are called ions. Ions can be individual atoms or they can be small groups of atoms.

ATOMS

H

BONDS

H–O–H

H_2O

MOLECULES

O–H
O=C
O–Na

$NaHCO_3$

IONS

O–H
O=C
O^- Na^+

Na^+
HCO_3^-

refers to whatever bits and pieces of materials chemists use. *Particle* is a similar term.

Atoms are the individual building blocks of elements. There are approximately 100 elements in the world; hence chemists work with approximately 100 different kinds of atoms. Symbols are used to represent atoms; these symbols are one or two letters taken from the names of the elements. In some cases, the symbol that represents an atom is derived from the element's name in a language other than modern English.

Molecules, which are made up of atoms, are the individual building blocks of *compounds*. *Formulas* are used to show the kinds and numbers of each element's atoms that have been put together to form a compound. Some compounds, particularly salts, are put together from pairs or groups of electrically charged fragments called *ions*, rather than from molecules. A grain of table salt, for example, is a crystal that is put together out of pairs of sodium and chloride ions. In this case, the formula, NaCl, represents one pair of these ions rather than one sodium atom and one chlorine atom joined together to form a molecule.

There are sharp differences between compounds which come apart to form molecules and those which form ions. Mixing solutions of substances which form ions creates opportunities for new combinations of ions to form quite rapidly. More strenuous methods must be used to cause two molecules to allow their atoms to be rearranged or reshuffled.

In this book, we use the word molecule to refer to the individual building blocks of nearly all kinds of materials. It is convenient to infer that putting together individual molecules is the goal of chemists. It is also convenient to infer that individual molecules show the properties or characteristics of compounds; that individual molecules of morphine kill pain, that the flavor of peppermint, the vivid colors of autumn leaves, the bounciness of rubber, or the pictures painted on TV screens can be traced to individual molecules.

Actually, chemists put together billions of molecules or ion pairs at a time. Similarly, the usefulness of a compound depends upon the combined effects of billions of molecules. It is true that chemists work with quantities of gases, liquids, and solids we can see. To appreciate how

Formulas represent the smallest possible particle of compounds. Small numbers below each symbol show the number of each kind of atom in the particle. Sometimes parentheses are used to simplify formulas.

FORMULAS

SIMPLIFYING FORMULAS

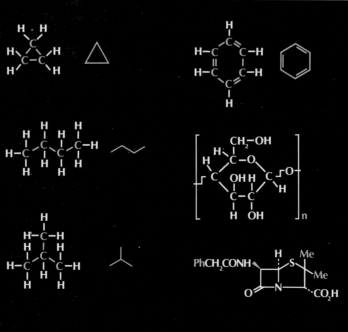

To simplify structural formulas, chemists have agreed that a carbon atom can be assumed to be at the end of each line that represents a valence bond unless another kind of atom is shown. Because each carbon atom usually has four valence bonds, hydrogen atoms attached to carbon atoms are sometimes omitted from structural formulas and it is up to the reader to realize where they belong.

Molecules can consist of atoms bonded or linked together like a chain. The chain can be either continuous or branched.

Molecules can also consist of atoms joined together to form a ring or "closed chain." Benzene is a very common 6-carbon atom ring with the equivalent of double bonds between every other carbon atom.

Sometimes chemists use the subscript "n" or "x" to show that a portion (monomer) of a large molecule is repeated n or x times to make a giant molecule or polymer.

Heavier lines are used when it is important to suggest the three-dimensional properties of a molecule.

Sometimes it is an advantage to abbreviate groups of atoms which occur in complex molecules. "Me" can represent a CH_3 or methyl group, "Et" can represent a C_2H_5 or ethyl group, or "Ph" a benzene ring or phenyl group, C_6H_5.

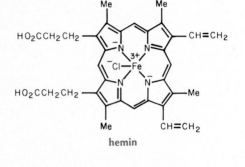

hemin

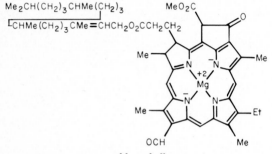

chlorophyll

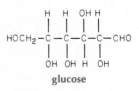

glucose

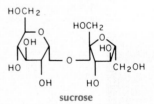

sucrose

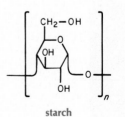

starch

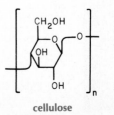

cellulose

Countless substances which build the tissues of plants and animals are also nature's chemicals

Nature's building blocks in rocks and soil, the air and water, are the same as those in the cells and tissues of living systems. The earth's elements are constantly being taken up and used to put together the molecules which build plant and animal bodies.

The uniqueness of each plant or animal suggests that there are as many different kinds of substances—if not more—as there are kinds of plants and animals in nature. Bone is but one example. The food we eat or drink and the air we breathe are not at all the hard, "bony" framework that supports and protects the other tissues of our bodies. By taking food apart, the body obtains the building blocks which can be put together to make bone.

Blood is another example. The blood flowing through our arteries and veins contains hemoglobin, along with other substances. Each red cell (and there are 220 million red cells in each drop of blood) contains as many as 333 million hemoglobin molecules. To put together a single hemoglobin molecule the body needs 2952 carbon atoms, 832 oxygen atoms, 4664 hydrogen, 812 nitrogen, 4 iron, and 8 sulfur atoms.

To add to our appreciation of the efficiency of nature's chemistry, a red cell must be replaced after only 120 days of existence. This means that all 333 million of its hemoglobin molecules must be put together every 120 days.

Only rarely are visible chunks of a pure element found in plant or animal tissue. In the non-living world, we find nuggets of pure gold or large chunks of copper. The air contains oxygen, nitrogen, and other elements which remain separate even though they are mixed thoroughly.

As we have already pointed out with hemoglobin, the molecules which form the cells and tissues of living systems usually require relatively large numbers of atoms in each molecule. A chlorophyll molecule, for example, calls for 55 carbon atoms, 72 hydrogen, 1 magnesium, 4 nitrogen, and 5 oxygen atoms. Ordinary table sugar, one of the simpler molecules of the plant world, calls for 12 carbon, 22 hydrogen, and 11 oxygen atoms in each molecule.

There is great variety in the properties of compounds

Although elements provide the needed building blocks, ours is a world primarily of compounds—a world of vitamins, fats, proteins, hormones, phosphates, fluorides, oxides, and on and on. This is why chemists devote so much time and energy trying to understand how molecules come apart and how the pieces can be put back together to form new kinds of molecules.

The proper kinds and numbers of atoms put together in the proper way will yield molecules with the sourness of vinegar or the sweetness of sugar, the flavor of wintergreen or the fragrance of a gardenia, the scent of a skunk or the bounce of rubber. Choose the right combinations of atoms and you will have molecules with every color imaginable; they will be as waterproof as paraffin or as water soluble as salt, as flammable as paper or as fireproof as granite, as transparent as glass or as opaque as asphalt.

Nature's building blocks can be rearranged when energy interacts with molecules

There is a very delicate balance between the kinds and numbers of atoms that stick together to form a molecule and the amount of energy that is available. Energy is also tied in with the motion of molecules; the more heat energy molecules absorb, the greater their motion. It follows that the greater the motion of molecules, the more likely they will bump into each other with enough impact to break the molecules apart.

Actually, energy's total role in taking things apart and putting things together is very complex. In one sense, energy can be the "glue" that holds atoms together to form molecules. At the same time energy can be responsible for causing molecular collisions which result in the shattering of molecules into fragments.

When collisions between molecules result in the molecules being broken into fragments, the fragments can join to form new kinds of molecules. Sometimes the new molecules capture additional energy. Sometimes while the fragmenting collisions are occurring, energy escapes from the system. Thus some chemical changes result in the storage of energy, whereas others result in the loss of energy. The terms, endothermic and exothermic, are used to distinguish between these two types of reactions.

Both kinds of reactions occur in nature. In a very important endothermic reaction, green plants absorb the sun's energy, and this energy becomes stored in energy-rich molecules of sugars and starches. In an equally important exothermic reaction, these molecules can be taken apart to yield energy-poor molecules of water and carbon dioxide, and the excess energy becomes available for the life processes and activities of living systems.

People gain much from nature's chemistry

Many of the materials we need for our comfort and well-being are created by the natural interactions between energy and matter in the environment. But people learned how to take lessons from nature's chemistry, to copy nature, so to speak, and use chemistry to produce things which make life more efficient.

Baking soda, $NaHCO_3$, can be used to illustrate the above ideas. Baking soda contributes to our comfort and well-being in many ways. Biscuits and other baked foods that are light and well leavened represent only one of the uses of baking soda. It can also be used to keep refrigerators sweet and clean, to brighten smiles and leave mouths feeling clean and fresh, and to make baths more refreshing. Many tons of this chemical are used each year in equally important but less familiar situations.

Nature's chemistry has produced large quantities of baking soda in various parts of the world. Near Green River in Wyoming, for example, there are large deposits of the mineral trona which contains baking soda. According to geologists, the story of this deposit of millions of tons of baking soda began some 50 million years ago when an immense lake, Lake Gosiute, covered southwestern Wyoming. Water from the surrounding area drained into Lake Gosiute bringing with it the minerals which dissolved as the water flowed through rocks and soil. Large populations of shell-forming animals flourished in the lake.

As time went on, the constant evaporation of water from the lake created a rich "soup" of dissolved minerals. But the climate changed drastically during Lake Gosiute's history. At times, the lake nearly dried up; at other times, the original water level was restored. Changes in the amount of water, together with changes in the amount of the sun's energy that fell on the lake, caused much taking apart and putting together among the minerals that were dissolved in the water.

At times, materials which had been dissolved could no longer stay in solution. Fragments from dissolved substances would combine and crystallize or settle to the bottom. During one exceptionally dry spell, thick piles of limestone, dolomite, and mudstone built up in the lake. The most soluble materials became concentrated in the water that remained in the deeper parts of Lake Gosiute.

Eventually all water evaporated. Even the most soluble minerals formed crystals. Trona—along with other minerals—accumulated in 10-ft thick beds. The atoms of the minerals originally washed into Lake Gosiute had been taken apart and the fragments rearranged to form new kinds of particles.

The beds of trona included baking soda, sodium bicarbonate, washing soda, sodium carbonate, Na_2CO_3, and several other kinds of material.

Energy plays not only an essential, but also a very complex, role in chemistry. Sometimes energy takes molecules apart and other times it is energy that causes atoms to join together to form molecules. It is also a question of whether energy is being built into or is escaping from the molecules.

Today, these beds are covered with some 1000 ft of rocks and soil that have been brought by erosion to where Lake Gosiute had been. The trona can be reached through mines, and provides a source of baking soda.

This baking soda was produced "as a matter of course" in nature. People can also struggle to get the things to happen that produce baking soda. But baking soda's story is somewhat different when it is made by people. One approach that people could use would be simply to track down the same raw materials and mix them together in the same way they were combined in nature. Sometimes this is how nature is copied.

There is another approach. It begins with trying to picture the kinds of building blocks needed and how they are put together to yield a desired product. In the case at hand, we know that individual particles of baking soda consist of pairs of sodium ions and bicarbonate ions. A sodium ion is a sodium atom which has lost one electron from the outer portion of the atom. Bicarbonate ions consist of one atom each of hydrogen and carbon, three oxygen atoms and a "borrowed" electron, borrowed perhaps from a sodium atom.

Another step is to identify a source of these two kinds of ions. Ordinary table salt, sodium chloride, is an abundant and cheap source of sodium ions. Bicarbonate ions can be obtained by bubbling carbon dioxide through water. Obtaining water poses no problem, and carbon dioxide is a by-product of burning coal, gas, or any other fossil fuel.

Now comes a much more difficult step: causing the sodium and bicarbonate ions to combine in the proper way to make baking soda or sodium bicarbonate. Many trials and errors had to take place before chemists solved this problem.

In 1886, a Belgian chemist, Ernest Solvay, came up with the idea of using an intermediate or go-between particle to cause the sodium and bicarbonate ions to combine in the proper way. Rather than trying to get these two kinds of ions to combine directly, he first added ammonium ions to the bicarbonate ions. This produced ammonium bicarbonate, a water-soluble compound which separates into ammonium and bicarbonate ions when it dissolves.

Salt, NaCl, is also highly soluble in water; it breaks up into sodium and chloride ions. When salt and ammonium bicarbonate solutions are mixed, the sodium and bicarbonate ions combine and settle out of solution. The resulting baking soda can be filtered off, dried, and packaged for sale.

We don't know if ammonium ions played a similar role when baking soda was being made in old Lake Gosiute. But we do know that Solvay's way of making baking soda is quite efficient. For example, since ammonia is expensive and only assists in making baking soda, it is recycled. To accomplish this, limestone is heated. This yields lime and carbon dioxide. (The carbon dioxide can be used to make carbonate ions.) When lime is added to the ammonium chloride solution remaining after the sodium bicarbonate has settled out, the calcium takes the place of the ammonium ions. The ammonium ions pick up electrons and bubble from the solution in the form of ammonia gas. The gas can be collected and recycled.

To further reduce the cost of making baking soda, the solution that is left—after the ammonia bubbles out—can be allowed to evaporate so that the calcium chloride can be harvested, dried, and sold as a by-product. Large quantities are used each winter, for example, to melt ice on streets and roads.

Chemistry is an invisible world of atoms and molecules

Many young people who are beginning their study of chemistry are puzzled when they hear that iron and gold, arsenic and oxygen, and all the other elements are made up of discrete but fantastically tiny particles called atoms. Their imaginations are stretched even further when they are told that all atoms are made up of electrons, protons, and neutrons. Then, of course, they want to know something about electrons, protons, and neutrons. At this point, they may be told that electrons are the tiniest possible bundles of negative electrical charge, protons are

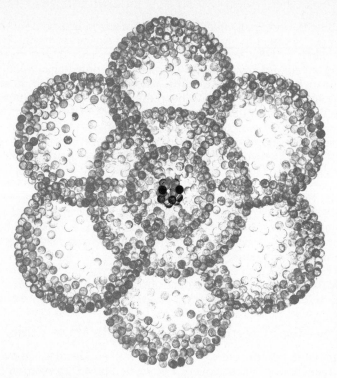

Although mental models of atoms always stretch our imaginations, we do know how many protons (red) and neutrons (purple) are in the nucleus and how many electrons (green) are built into each kind of atom.

equally tiny bundles of positive electrical charge together with something else that makes them weigh nearly 2000 times as much as electrons, and neutrons are particles which lack electrical charge but weigh about the same as electrons and protons combined.

These ideas become even more difficult to accept when the young people are told that the only differences between the atoms of the various elements are differences in the number and arrangement of electrons, protons, and neutrons in their atoms. Typical reactions are, "Prove it," or "So what," or "Does this mean I can crunch a lot of iron atoms together to make gold?"

It is to be expected that students should react this way and raise the questions they do. It is difficult to answer such questions, however, without repeating the experiments which enabled scientists to prove for themselves that electrons, protons, and neutrons exist and play major roles in putting atoms together. Until people have worked with equipment and apparatus and have the experience that comes with research, in the words of one young person who was especially puzzled by atomic theories, "I guess we will have to take the scientists' word for it."

Sooner or later, however, people who hope to appreciate or work with chemistry need to put together their own mental model of atoms. It is hazardous to say that one model would be better than another. Brains are terribly private. The mental model of atoms that "works" for one person may be most misleading to another. The important thing seems to be to put together the best model possible and then be ready to redesign the model when new facts and insight come to mind. Here is an example of one person's mental model of atoms.

An atom is like the shimmering, transparent disc that appears around the blades of a whirling fan or spinning airplane propeller. The protons and neutrons are packed together at the hub and form only a very, very small part of the total volume of the atom. Most of the atom is the imaginary, shimmering illusion. It exists only because the electrons passed by recently and are sure to return.

In contrast to the disc created by the blades of a propeller, the electrons of an atom create a spherical domain. The positively charged nucleus keeps the electrons from breaking loose from the shimmering sphere just as the hub keeps the blades from flying off. One atom differs from another only because there are more particles packed into the nucleus and more electrons whirling and spinning. Two or more atoms could bond or lock together if their spinning electrons were precisely synchronized. The similarly charged nuclei, of course, would repel each other and keep bonded atoms from coming together completely.

This mental model may or may not come close to being what nature's building blocks are actually like. It is, at least, a working model for the person whose model it is, and it is always fun to exchange models with other people.

Civilization had not advanced very far before people began asking "how" and "why" questions about the materials which helped them survive. Especially puzzling must have been the fact that the properties of a substance could be so different from the properties of the raw materials which produced it. It wasn't long before people began to dream of putting together materials with all manner of new and wonderful properties—materials that would change cheaper metals to gold, for example, or the "philosopher's stone" or the "elixir of life."

Obviously, the grand dreams of the alchemists were never realized. They never found the stone that would change lead to gold or the elixir that would bring a life of eternal happiness. But what came from their searching may have turned out to be almost as valuable. This was the realization that there was a world of building blocks and that things could be taken apart, things could be put together.

To discover a new element called for men and women who were brave enough to "see" in the properties of a substance a not-yet-known element and to follow the clues their observations and hunches provided.

Polonium—the element that sent signals to Marie Curie

It is difficult to say for sure what first triggered an idea that led eventually to an invention or a discovery. But we do know that even "accidental" discoveries are most likely to be made by people who have some kind of investigation under way. Similarly, the idea for a new investigation often is triggered by something that the investigator learns while pursuing an earlier investigation.

Near the end of the 1800s, Marie Curie was in Paris studying the magnetic properties of iron and steel. In 1898, her husband, Pierre Curie, and his brother, Jacques, discovered piezoelectricity. This is the property of certain kinds of crystals to produce an electrical charge when the crystals are squeezed. Part of their work called for the invention of a meter that would detect very weak electrical currents.

Two years earlier, Henri Bequerel had announced the discovery that certain minerals, especially those containing uranium, would expose photographic film even though the film was wrapped in lightproof paper. The announcement of this very puzzling discovery could well have been a part of the reason for Marie Curie to drop her investigation of magnetism and take up the investigation of the strange behavior of uranium minerals.

One of the things she did was to see if uranium minerals caused any effects on the delicate instrument that had been invented by her husband. She found that something that was given off by the minerals could be counteracted by the electricity produced by squeezing crystals. In time she stated that the compounds of uranium and thorium emitted rays which exposed photographic film and caused air to conduct electricity even though the rays couldn't be seen, felt, or detected in other ways.

Marie Curie became particularly puzzled by pitchblende and chalcolite, two minerals that contain uranium. Sometimes samples of these minerals gave off more rays than could be accounted for on the basis of the amount of uranium they contained. At this point, she decided that these mineral samples might contain a not yet discovered element that was also radioactive, the term she introduced to account for the puzzling properties of uranium and similar elements.

How Madame Curie went about proving her hypothesis by obtaining a sample of this new element illustrates beautifully how this kind of taking apart and putting together is done. Fortunately, pitchblende could be obtained in large quantities, tons, that is, at low cost. The usual methods were used to crush and grind the mineral so that it could be treated with acids. The argument was that the elements known for sure to be in the pitchblende could be dissolved, converted to insoluble compounds which would precipitate from the solution, and filtered off and discarded.

Hunting for this new element had one advantage not shared by those who discovered many of the other elements. It was easy to determine where the target was, that is, to know whether the new element stayed in

Alpha Particle

Beta Particle

Gamma Rays

Radium

Marie Curie

The great achievements of Madame Curie bear witness to the effectiveness of a brilliant mind aided by precise instruments and driven by the burning desire to understand our universe.

the solution or left with the precipitate. Her husband's sensitive electrometer would detect this. In fact, by now her husband had joined in the search for the new element.

After using their knowledge of chemistry to select the kinds of acids and other reagents that were known to precipitate all of the elements that were known to be in pitchblende, the Curies were left with a tiny amount of material they knew contained bismuth hydroxide, $Bi(OH)_2$, but which also showed the property of radioactivity. And bismuth hydroxide couldn't be responsible for the radioactivity.

Finally, the Curies found that if sulfide ions were added and the resulting black precipitate heated to 700°C in an evacuated tube, a black deposit appeared on the walls of the tube. This black deposit proved to be 400 times more radioactive than uranium. Quoting the Curies, "We believe, therefore, that the substance we have isolated from pitchblende contains a heretofore unknown metal. . . . If the existence of this new metal is confirmed, we propose to call it *polonium*, after the name of the native country of one of us (Madame Curie)."

Actually, the substance which collected on the walls of that test tube turned out to be doubly rewarding. By continuing to dissolve this substance and then adding other chemicals that would precipitate the metals which were known to be present, they discovered a new metal—more than 900 times as radioactive as uranium. Now known as radium, this element has found even more uses than polonium.

The intense motivation and determination of the Curies have been widely dramatized—and rightly so when we realize that 10 million grams of pitchblende must be taken apart to obtain one gram of radium.

The discovery and commercial production of aluminum

Sometimes a newly discovered element calls for additional research before it can be extracted from its sources in quantities large enough to supply the demand. Aluminum is a good example.

Aluminum's story began during the 18th and 19th centuries when many people were interested in all kinds of rocks and minerals. Part of this interest involved elements each mineral contained. After discovering that water is put together from hydrogen and oxygen, the French chemist Lavoisier tried to extract a metal he believed existed in clay. His work on this project, however, ended when he died on the guillotine during the French Revolution.

Sir Humphry Davy and H. C. Oersted transferred their experience, using electricity to extract metals from minerals, to the problem of extracting an unknown metal from clay. They caught glimpses of a metal which in color and luster resembled tin.

Two years later, in 1845, Frederick Wöhler and Sainte-Claire Deville improved on the idea attempted by Davy and Oersted. They converted the as-yet-unknown metal to its chloride and then heated the chloride with potassium or sodium dissolved in mercury. Soon bits of the new metal, aluminum, "as big as pinheads" appeared and grew to "lumps the size of marbles."

The aluminum obtained by this process, although very expensive, revealed its useful properties. Factories were built to produce aluminum commercially. At this time, Charles Martin Hall, while a student at Ohio's Oberlin College, was inspired by a chemistry professor to start looking for a cheaper way to extract aluminum from its ores.

Perhaps it was by looking back at earlier attempts to produce aluminum that Hall got the idea that electricity could be used to separate aluminum from its ores. But bauxite, the best aluminum ore, does not conduct electricity. Hall knew, however, that many substances became conductors when dissolved in water or heated hot enough to melt.

Since bauxite does not dissolve in water and could not be melted economically, Hall looked for some solvent that would dissolve bauxite and yield a conducting solution. Eventually he tried cryolite, Na_3AlF_6. If cryolite is heated to 982°C, it melts, and the molten cryolite dissolves bauxite. Now the aluminum can be extracted by passing electricity through the cryolite-bauxite solution.

In actual practice, the bauxite is first ground to a powder and con-

Atoms: building blocks of the universe

To group the earth's elements by bringing together those which share similar properties helps us not only to better understand their chemistry but also to predict how they can be put together to make the things we need.

verted to pure aluminum oxide, Al_2O_3. The aluminum oxide dissolves in molten cryolite better than does crude bauxite. Much of Hall's success was due to his idea to line the crucible in which the electrolysis was carried out with carbon. The carbon lining wore away but it lasted long enough to make the process a commercial success.

Paul Heroult, a French chemist, came upon almost the same way to produce aluminum while he was looking for ways to improve electric furnaces. His discovery was made at the same time as Hall's; therefore, the molten cryolite method for producing aluminum is called the Hall-Heroult process.

The advancing stream of civilization is pretty much a story of turning to the earth's rocks and soil, its air and water for the raw materials we need to solve our problems and fulfill our needs. The advance of civilization depends upon maintaining an adequate supply of the earth's building blocks. This means people must know where supplies of the elements can be found and how to harvest and recycle what they need.

Where a particular element can be found depends upon its properties. Substances which boil at temperatures below that of the environment, for

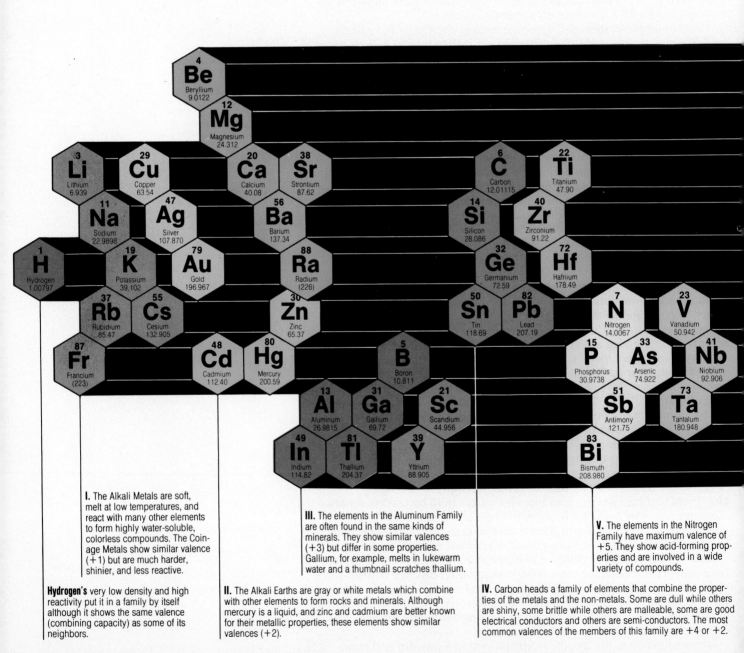

I. The Alkali Metals are soft, melt at low temperatures, and react with many other elements to form highly water-soluble, colorless compounds. The Coinage Metals show similar valence (+1) but are much harder, shinier, and less reactive.

Hydrogen's very low density and high reactivity put it in a family by itself although it shows the same valence (combining capacity) as some of its neighbors.

II. The Alkali Earths are gray or white metals which combine with other elements to form rocks and minerals. Although mercury is a liquid, and zinc and cadmium are better known for their metallic properties, these elements show similar valences (+2).

III. The elements in the Aluminum Family are often found in the same kinds of minerals. They show similar valences (+3) but differ in some properties. Gallium, for example, melts in lukewarm water and a thumbnail scratches thallium.

IV. Carbon heads a family of elements that combine the properties of the metals and the non-metals. Some are dull while others are shiny, some brittle while others are malleable, some are good electrical conductors and others are semi-conductors. The most common valences of the members of this family are +4 or +2.

V. The elements in the Nitrogen Family have maximum valence of +5. They show acid-forming properties and are involved in a wide variety of compounds.

example, exist as gases in the atmosphere. Exceptions are quantities of gases that have been trapped in porous rocks or dissolved in water or petroleum. Substances which dissolve in water exist in solutions unless they are deep underground or otherwise kept away from water.

Solids are substances which freeze at temperatures above those of the environment and are to be found in rocks and soil.

Water shows how temperatures influence the state in which a substance is to be found. In arctic regions where temperatures stay below freezing, water exists as solid ice and snow. Where environmental temperatures are above freezing, water maintains an equilibrium between liquid and gas, but if temperatures were to rise above the boiling temperature, all water would become clouds of steam. It is interesting to realize that no other substance is as familiar in all three states as water and that it is liquid water that supports living systems.

The earth provides approximately 100 elements, "approximately" not because chemists haven't counted the elements but because of the expression, "the earth provides." Eighty-eight elements have been identified in the earth's rocks and soil, air and water. Eleven additional elements are

2 He Helium 4.0026		57 La Lanthanum 138.91		
9 F Fluorine 18.9984	17 Cl Chlorine 35.453	10 Ne Neon 20.183	58 Ce Cerium 140.12	59 Pr Praseodymium 140.907	89 Ac Actinium (227)
35 Br Bromine 79.909	53 I Iodine 126.904	25 Mn Manganese 54.938	18 Ar Argon 39.948	60 Nd Neodymium 144.24	61 Pm Promethium (147)	90 Th Thorium 232.038	91 Pa Protactinium (231)
85 At Astatine (210)	43 Tc Technetium (98)	75 Re Rhenium 186.2	36 Kr Krypton 83.80	54 Xe Xenon 131.30	62 Sm Samarium 150.35	92 U Uranium 238.03
8 O Oxygen 15.9994				86 Rn Radon (222)	63 Eu Europium 151.96	93 Np Neptunium (237)	94 Pu Plutonium (242)
16 S Sulfur 32.064	24 Cr Chromium 51.996			26 Fe Iron 55.847	64 Gd Gadolinium 157.25	95 Am Americium (243)
34 Se Selenium 78.96	42 Mo Molybdenum 95.94			27 Co Cobalt 58.933	65 Tb Terbium 158.924	96 Cm Curium (247)
52 Te Tellurium 127.60	74 W Tungsten 183.85			28 Ni Nickel 58.71	66 Dy Dysprosium 162.50	97 Bk Berkelium (247)
84 Po Polonium (210)			44 Ru Ruthenium 101.07	67 Ho Holmium 164.930	98 Cf Californium (249)	99 Es Einsteinium (254)
		45 Rh Rhodium 102.905	68 Er Erbium 167.26	100 Fm Fermium (253)
		46 Pd Palladium 106.4	69 Tm Thulium 168.934	101 Md Mendelevium (256)	102 No Nobelium (254)
77 Ir Iridium 192.2	76 Os Osmium 190.2	70 Yb Ytterbium 173.04	103 Lw Lawrencium (257)
78 Pt Platinum 195.09	71 Lu Lutetium 174.97

VI. The Oxygen and Sulfur Family includes well-known as well as less common elements. Although the members of this family differ in many ways from the metals, some well-known metals show some of the same chemical properties.

VII. The Halogen or "Salt-Forming" Family of elements combine readily with other elements to form compounds that are usually colorless and water soluble.

VIII. The Noble Gases find uses in advertising signs and for filling balloons, but their most distinguishable property is that they do not combine (valence 0) at all readily with other elements.

The Actinide Family contains elements which have been featured in recent worldwide events, as well as elements which have been created only in scientific laboratories.

The Rare Earths is a family of closely related elements that have been difficult to separate and identify. Interesting uses have been found for some of these elements. Cerium, for example, helps create the shower of sparks which ignite cigar and cigarette lighters.

known, but people have had to give nature a hand in order for some of these to exist. These are the transuranium elements, elements whose atoms have been created by firing fragments of atoms into the nuclei of some of the heaviest naturally occurring atoms. For each additional proton that is made to lodge in a target nucleus, a new element is created.

The elements provide a wide array of properties. Some can be as poisonous as plutonium or as life-saving as oxygen, as hard as tungsten or as soft as potassium, as colorful as bromine or as colorless as neon. Some elements are as brittle as manganese or as malleable as gold, as dense as uranium or as light as hydrogen, as shiny as silver or as dull as sulfur, as ductile as copper or as crumbly as carbon.

Some elements are good electrical conductors, whereas others are equally good insulators. The protons and neutrons of most atoms stay put in the nucleus but atoms of the radioactive elements, sooner or later, are sure to lose or rearrange one or more of the particles in their nuclei.

The elements also differ in how actively their atoms bond or stick to other kinds of atoms. Helium, neon, and the other noble gases form compounds with other elements only under carefully managed conditions. Gold and platinum combine with other elements only at high temperatures or when catalysts or strong acids are involved. For other elements, their atoms are so reactive that these elements exist in nature only in compounds.

Much of the action in chemistry is in arranging and rearranging atoms

Discovering the elements, becoming familiar with their properties, and finding ways to extract them from their sources has been a great chapter in the total story of chemistry. The main action nowadays, however, has shifted to finding how to build atoms into new kinds of molecules—new molecules which will help us understand our world, fulfill our needs and solve today's problems.

The properties of a substance are locked into the kinds, numbers, and arrangement of the atoms in its molecules or ion pairs in its crystals. Change these building blocks and we make a new substance. Its properties will be different and we can always hope that the new properties will answer our questions or fulfill our needs.

The work of Nathan B. Eddy and Everette L. May provides a good example of this kind of chemistry. They worked with morphine, a substance that relieves pain—but it is not a perfect medicine. The dose must be controlled carefully, and when it is used repeatedly, the dose must be increased. Worst of all, morphine causes addiction.

Morphine, as morphine, has been known for less than 200 years. But the source of morphine has been known for more than three thousand years. Longer ago than that children were given poppy seeds to chew to keep them quiet. Opium was found to be the substance in the poppy seeds that quieted the children. For centuries opium in one form or another was in every medical kit. It was used as a household remedy for mental disorders and for aches and pains of all kinds.

Then people began wondering what there was in the opium that gave it its properties. By 1803, chemists had extracted from opium a crystalline substance that was responsible for opium's effects. It was named morphine after Morpheus, the god of sleep.

Chemists learned to separate morphine from all the other substances in opium in the hope that the pure substance would not have opium's harmful side effects. Other chemists worked out the kinds and numbers of atoms in morphine molecules and how these atoms were arranged.

By 1847, it was known that each morphine molecule was built from 17 carbon atoms, 19 hydrogen, one nitrogen, and three oxygen atoms. By 1925, chemists knew how all of these atoms fit together to form the morphine molecule. During all these years, morphine continued to be used as an essential medicine.

In order to work out how the morphine molecule was put together, chemists treated morphine with various chemicals. These treatments changed the morphine molecule. One such change produced heroin. In 1898, heroin was introduced as a medicine that was thought to be su-

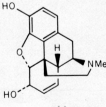

morphine

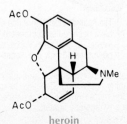

heroin

perior to morphine. It seemed to be safer when used to relieve pain, and when first introduced, it was reported to be non-addictive.

As more and more case histories were reported, however, heroin was found to be more of a problem than morphine. Its side effects were bad and it turned out to be addictive. This illustrated an important fact about complex molecules. One portion of the molecule may be responsible for one of the properties of the substance.

It was this realization that prompted Dr. May to see if he could determine exactly what part of the morphine molecule was responsible for relieving pain and what portion caused addiction. He believed that he might be able to redesign the morphine molecule so as to keep the pain-relieving property and get rid of its bad effects.

Intrigued by this challenge, Dr. May and other chemists have devoted many years to trying to redesign the morphine molecule. In general, there has been, at best, limited success. It may be that addiction to morphine or heroin is a behavior that is only partially related to the actual chemistry of the drug. But as this kind of research is continued, sooner or later someone will very probably design the near-ideal substitute for one of nature's very useful yet dangerous medicines.

It may seem strange to include in this book the story of a not-yet-successful project. There are hundreds of projects in which chemists have succeeded in remodeling nature's molecules to improve their effectiveness in solving our problems and fulfilling our needs, and these stories are fun to tell. Were we to include only those stories in which men and women have scored successes, however, we would be like a football commentator who talks about only the player who happens to be carrying the ball when a touchdown is scored.

Although chemists have not yet succeeded in redesigning the morphine molecule in a way that will remove the addiction property, they have certainly scored a "touchdown" in a closely related project. Until early in the 1900s, there was no anesthetic that could be used to control the pain that accompanies dental work except alcohol and cocaine. Both of these substances have dangerous side effects, especially cocaine.

In 1906, a United States patent was issued for a new substance that has proved to be of enormous help in controlling the pain accompanying dental work. Dr. H. Braun, a German chemist, found that the substance procaine hydrochloride, sometimes known as novocaine, deadened pain. He proved this by injecting dilute solutions of the drug into his fingers. His fingers became anesthetized immediately, but sensitivity returned within a few minutes. It was later found that a very powerful anesthetic could be prepared by adding epinephrine to the procaine hydrochloride. The side effects of this drug cause very few problems among dental patients.

Another story involving anesthetics illustrates how discoveries sometimes happen when men and women are looking for better solutions to problems. A worldwide search for a general anesthetic more suitable than ether, chloroform, or nitrous oxide was reaching its peak about 1923. As a part of this search, Drs. V. E. Henderson and W. Easson Brown at the University of Toronto looked into the use of propylene as an anesthetic. They abandoned their project, however, when they found that the propylene seemed to pick up toxic effects when used as an anesthetic under certain circumstances.

When George H. W. Lucas became a colleague of Drs. Henderson and Brown, they suggested that he might throw some light on the nature of the toxic material which seemed to occur in propylene. This suggestion arose at least partly because a tank of propylene remained in their laboratory as "mute evidence" of their failure to find a suitable new anesthetic.

Although he does not explain exactly how it happened, Dr. Lucas found that the propylene molecules if left standing long enough underwent "natural" rearrangement of their atoms. Rather than having the three carbon atoms share one single and one double bond and stay in a chain-like molecule, Dr. Lucas hypothesized that the chain doubled around, so to speak, to form a closed ring. Such a compound is cyclopropane.

Acting on the hunch that cyclopropane might be a good anesthetic, Dr. Lucas prepared several liters of pure cyclopropane. He then put two small kittens in a bell jar and filled the bell jar with a mixture of cyclopropane and oxygen. The kittens went to sleep quietly with "no toxic

coacaine

procaine hydrochloride

$H_2C = CHMe$
propylene

Et_2O
ether

cyclopropane

$H_2C = CH_2$
ethylene

manifestations." They recovered promptly when removed from the bell jar. When the experiment was repeated later in the day, again the kittens went to sleep quietly and recovered nicely when taken from the bell jar.

Further studies proved that cyclopropane worked very well when used as an anesthetic for cats and rabbits during surgery. Looking ahead to using this anesthetic on people, Dr. Lucas also investigated its flammability when mixed with oxygen. In one experiment, he prepared a 100-milliliter mixture consisting of 16 percent cyclopropane in oxygen. After wrapping the container with copper screen, Dr. Lucas crawled under a laboratory table before striking an electrical spark to the mixture. The resulting explosion was best described by quoting Dr. Lucas' lab report, "Blew all to hell."

Dr. Henderson was the first person to volunteer to test the effects of cyclopropane as an anesthetic for people. Other trials followed promptly, and within a few years cyclopropane began replacing ethylene to the satisfaction of anesthetists, surgeons, and patients. By 1934 cyclopropane was chosen by many surgeons in preference to ether in well over 75 percent of the work formerly done with ether.

It would be hazardous to suggest what the limits are to improving nature's chemicals. Very probably, the alchemists were asking too much when they tried to make their "philosopher's stone" or "elixir of life." It is highly improbable that anyone will ever come up with a substance that changes cheaper metals to gold or brings untold wealth, health and happiness.

In general, chemists choose goals that are worthy of their abilities and stand a chance of being achieved. Such choices are highly personal. Quoting Albert Szent-Gyorgi, one of America's great chemists, "As for myself, I like only basic problems, and could characterize my own research by telling you that, when I settled in Woods Hole and took up fishing, I always used an enormous hook. I was convinced I would catch nothing anyway, and I thought it much more exciting not to catch a big fish than not to catch a small one."

Other chemists like to set as their goal the solution to an immediate or practical problem. It can be the elimination of a simple nuisance or the prevention of a major catastrophe.

It is the role of chemistry to realize what properties a substance must have to solve a problem or fulfill a need. If the substance can be obtained from nature, the next step is to extract or separate the substance from its source. If it is a mineral, for example, it must be separated from other minerals; if a plant or animal product, the desired substance must be separated from all other substances in the cells and tissues of the living source. Finally, the substance must be purified, packaged, and made available in the quantities needed.

Sometimes nature's supply of a material is inadequate. Some natural products are very expensive to harvest, or they may have to be transported long distances. With the world population increasing as it is, the demand for some substances which once seemed to be only a small fraction of the supply has increased sharply. With increased demand, the natural sources are threatened to the point of extinction.

Ivory has the properties needed to make such things as piano keys, billiard balls, and handles for toilet articles. But the source of ivory is elephant tusks, and the only way to harvest ivory is to kill elephants. Tortoise shell makes attractive eyeglass frames and combs. But turtles must be killed to obtain tortoise shell. It is obvious where all of this leads.

The earth's rocks and soil, its atmosphere and hydrosphere, are enormous reservoirs of elements and compounds. If managed wisely and recycled efficiently, there are enough building blocks to fulfill our needs and solve our problems for many, many years.

But the advance of civilization takes its toll. Already we have harvested many of the richest, most easily obtained deposits of some materials. Oil and gas are worrisome examples. We need to take a fresh look at our trash heaps, our spoil banks, and our natural material and energy sinks and give nature a stronger hand at maintaining a continuous supply of the building blocks we need.

We must also give the sun a stronger hand at keeping us supplied with energy. After all, if the alchemists could dream of eternal health and happiness, should we be put down for dreaming of unlimited energy?

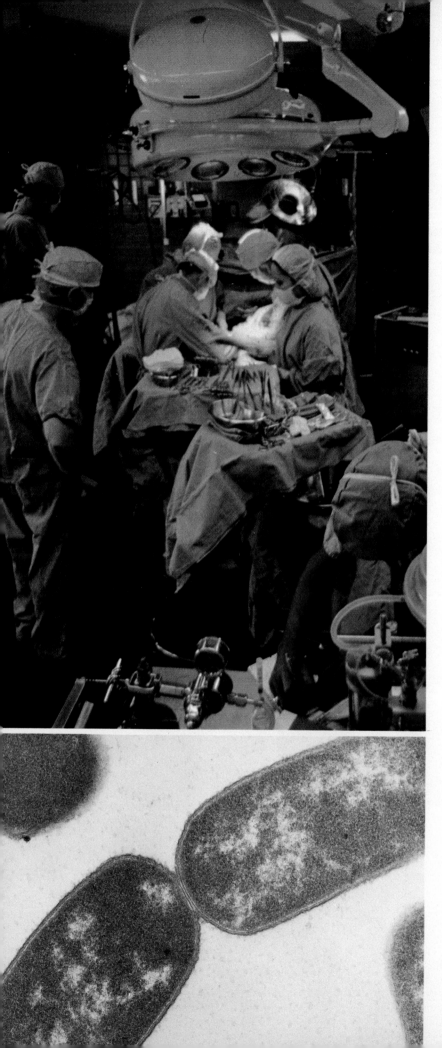

2
Inside Us:
The Chemistry
of Staying
Well

Other elements

Phosphorus

Calcium

Nitrogen

Hydrogen

Carbon

Oxygen

No matter how great are the achievements of our civilization, we cannot forget that it all begins with an intricate assembly of nature's building blocks.

Good health helps us do the things we like and need to do. Good health is natural but not automatic. Each year an outrageous number of people are besieged by disease or serious injury.

Today, one of every 50 babies born in the United States dies before becoming one year old. The infant mortality rate is higher in the U.S. than in 14 other countries. This is not necessary. Things can be done to improve the health of people. As evidence, 40 years ago one of every 200 mothers died during childbirth. This was reduced to one of every 500 by 1945, and today only one of every 3000 mothers dies during childbirth.

One hundred years ago, people in the U.S. who reached the age of 15 could look forward to living only 45 years more. Today, a 15-year-old person can expect to live at least 57 years more.

Changes in why people die provide more evidence that things can be done to improve health. Influenza, pneumonia, tuberculosis, and diphtheria, for example, were leading causes of death in 1900. Today, these diseases scarcely "make the charts."

There are people who believe that more of the world's people can have good health. These are the people who say, "If fewer babies die in one part of the world than in others, I've got to do something." These people deserve our help.

Good health depends on an intricate set of fantastically complex interdependent chemical reactions

Only four elements—carbon, oxygen, nitrogen, and hydrogen—provide the building blocks which create almost 99% of our body cells and tissues. Seven other elements account for almost all of the remaining 1%. There are 13 other elements that play vital roles but are present in only "trace" quantities. Iron, for example, is thought of as being a trace element even though four iron atoms are built into each hemoglobin molecule.

Atoms of these elements combine to form an enormous number of different kinds of molecules, all of which are involved in the chemistry that goes on in our bodies. Furthermore, there are many kinds of "in between" or activated complex particles which can be detected by instruments but which exist only momentarily when other kinds of molecules are being taken apart or put together. These particles exist long enough, however, to play important roles in health and disease.

Disease seldom results from only one system's being out of balance. Similarly, good health requires all body systems with all different kinds of molecules to be working together well. Furthermore, the body must be able to cope with dozens of kinds of "evil" or disease-producing molecules—molecules produced by mysterious microorganisms, or something as everyday as poison ivy.

"Raw material" building blocks must be available always

In the dynamic, ever-changing world of cells and tissues, food provides the building blocks needed for growth and for repair and replacement of "worn" molecules. The body also needs new cells and tissues to repair the damages of disease and injury. Much of the story of the growth and replacement of body tissues is the breaking apart of the proteins in food and the use of these fragments to put together the proteins needed to build and rebuild new cells and tissues.

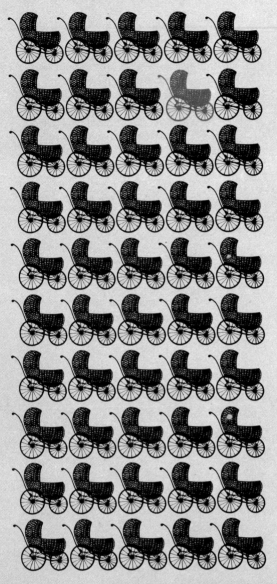

To reduce the number of babies who die in the United States each year can well be a challenge to tomorrow's scientists.

A fundamental feature of life is the interaction of energy with a specific array of essential elements whereby unique molecules are brought together in fantastically complex systems.

Atoms are brought together to form small molecules, amino acids for example. Small molecules are built into large molecules, proteins for example. Large molecules are built into cells, tissues, and organs.

Proteins are complex molecules put together from about 20 simpler molecules, the amino acids. Plants can build some amino acids from the nutrients absorbed from soil and air. Other amino acids are the products of animal metabolism.

Each living system involves an almost unique set of proteins. And each protein calls for a specific number of certain amino acid building blocks arranged or joined together in a precise way. There can be many different kinds of proteins. Suppose, for example, that we have five each of the 20 amino acid molecules and are going to put the 100 molecules together to form a protein. This means that 20 multiplied by itself 100 times is the number of different proteins that could be built from the 100 amino acid molecules. Someone calculated that if we had only one molecule of each of these different proteins, the total mass would be greater than the whole universe.

Actually, between 5 and 10 thousand kinds of protein molecules constitute all the protein content of the different cells and tissues in the whole plant and animal world. Good health for any plant or animal, however, depends on adequate quantities of a precisely established array of specific proteins.

The story of insulin

Insulin illustrates beautifully how a molecule, if available in adequate amounts, brings good health. The story of insulin also has a happy ending, although not as happy as many people would like.

The story begins for thousands of people when their doctors identify symptoms of weakness, weight loss, excessive appetite and thirst, and the excretion of excessive amounts of urine. If abnormally high amounts of sugar are found in the blood, and overflow into the urine, these additional symptoms suggest sharply that an upset has occurred in the body's production of insulin. The common name for this condition is diabetes.

As early as 1776 it was known that the pancreas, a large gland near the small intestine, had much to do with the body's ability to use sugar. By 1876, it had been proved that removing an animal's pancreas caused the animal to die regardless of how much food the animal ate. Finding the specific substance produced by the pancreas and enabling the body to metabolize sugar became a challenge to medical and chemical research.

Success came in 1922 in the laboratory of Dr. James MacLeod at the University of Toronto where Dr. Frederick Banting and Dr. Charles Best had been working for many years. The research they completed began many years earlier. In 1889, J. Von Mering and O. Minkowski proved that an animal whose pancreas had been removed would develop diabetes. In 1908, G. Zuelzer and E. Gley looked into the effects caused by closing the veins leading from the pancreas. They found that when the veins were tied shut or plugged, animals did not develop diabetes. The disease did not appear even after the lack of blood reduced the pancreas to a mass of fibrous tissue.

This led to the conclusion that only certain portions of the pancreas—portions now known as the islets of Langerhans—actually secreted the required substance.

Banting and Best then tied the ducts leading from a dog's pancreas. Ten weeks later, they removed the pancreas and chopped it up in an ice-cold solution that matched the fluids that surrounded living cells and tissues. In time, something had diffused from the chopped pancreas which, when injected into dogs whose pancreas had been removed, kept the dogs from becoming diabetic. Additional research proved that this substance, now known as insulin, could also be obtained from the pancreases of cattle. By now, Banting and Best knew they had a substance which could be injected into people whose pancreases did not produce the insulin they needed to "use" sugar.

There was the hope, however, that a less expensive and more easily controlled source of insulin could be found. By 1926, the pure substance responsible for insulin's effects had been isolated. It proved to be a protein. Now came the challenge of putting together this protein without relying on animals' pancreases collected from slaughtering plants.

With the use of methods with which chemists are familiar, the insulin molecule was found to be about 6000 times heavier than the lightest atom, hydrogen. Typical amino acid molecules are about 125 times heavier than hydrogen atoms. On this basis, the insulin molecule seemed to call for about 50 amino acid building blocks.

The next problem was to find the kinds of amino acid molecules that combine to form the insulin molecule. Studies involving the digestion of proteins have revealed that certain substances called enzymes help break proteins down into amino acids. Furthermore, each kind of enzyme acts most directly on certain amino acids. There is an enzyme, for example, that attacks the amino acid cysteine. When the insulin molecule was treated with this enzyme, the molecule broke into two pieces. Cysteine is the only amino acid with a sulfur atom in its molecule, and it was known that cysteine molecules can be linked together by sharing sulfur atoms.

Because the enzyme that breaks the bond between cysteine molecules also breaks the insulin molecule into two chunks, it is argued that the insulin molecule consists of two chains of amino acid building blocks joined by two cysteine molecules. By continuing this kind of strategy, it was proved that the insulin molecule consists of one chain of 30 and another of 21 amino acid units. Similar strategies were used to determine the exact sequence of each kind of amino acid in each chain.

$MeCH(NH_2)CO_2H$

alanine

$HSCH_2CH(NH_2)CO_2H$

cysteine

21

Finding how the insulin molecule is put together was a giant step toward the ending of the insulin story everybody is waiting for—the synthesis of a form of insulin that everybody whose pancreas doesn't work properly can swallow rather than inject. Or it may be that a different kind of molecule will be put together—one that will do what the insulin molecule does, and can be taken orally. Available today are preparations which seem to be effective for some people. The way the chemical profession and industry work, a happier ending to the insulin story may appear even before this book is finished.

Too much or too little of only one amino acid can upset many body systems

This is a story of symptoms as different as pale skin and faded hair, an unpleasantly pungent body odor, severe hyperactivity, impulsive self-injury, severe impairment of the total muscular system, and almost complete lack of intelligence. Fortunately, this story has a happy ending.

In 1934, a Norwegian mother sought help from Dr. Asbjorn Folling because she realized that her young son and daughter were becoming increasingly mentally retarded. Dr. Folling was particularly puzzled by the strange body odor of the children, an odor that was also associated with their urine. He used the usual solution of ferric chloride, $FeCl_3$, to test their urine to see if they were diabetic. Rather than the red brown color indicating diabetes, he found a puzzling brighter green color.

This strange observation caused Dr. Folling to look into everything that might have caused the children's urine to be unlike that of other people. Eventually he found a substance in their urine that seemed to cause the green color. His next step was to find out the chemical composition of this substance—how the molecules were put together.

When a bit of the substance was burned, only carbon dioxide and water were produced. This told Dr. Folling that the substance was put together from carbon, hydrogen, and, possibly, oxygen atoms. By carefully weighing a sample of the substance and the yield of water and carbon dioxide from burning the sample, he found that each molecule contained nine carbon, eight hydrogen, and three oxygen atoms. The formula for each molecule can be shown as $C_9H_8O_3$.

The next step was to find out how these 20 atoms were arranged in the molecule. To do this, Dr. Folling treated the substance with chemicals known to break the molecule into pieces he could identify. Eventually he decided that he was working with a substance called phenylpyruvic acid. To verify his hunch, he compared a sample of the substance obtained from the children's urine with a pure sample of phenylpyruvic acid to see if they had the same properties, especially the same melting temperature. They did.

Dr. Folling located eight other children, including two sets of siblings, whose urine showed the same green color when tested with ferric chloride. He was pretty sure he was dealing with an inherited inability of the body to handle a specific amino acid, phenylalanine.

Further study suggested that the inability to handle phenylalanine was passed from parents to their children by a single pair of recessive genes. In effect, the condition, now known as phenylketonuria and nicknamed the "green diaper" disease, was caused by the absence of a single gene. This gene, when present, caused the liver to produce phenylalanine hydroxylase, the enzyme required to take apart phenylalanine. If either parent gives this gene to a child, no symptoms of phenylketonuria appear.

Fortunately, the body's need to handle phenylalanine does not develop until several weeks after birth. This allows time for the "green diaper" test. Most encouraging of all, for those parents who learn that their child cannot metabolize phenylalanine, diets can be arranged which almost totally eliminate this amino acid. Dr. Frank L. Lyman, one of the people who has devoted much time to this disease, includes in his book, "Phenylketonurias," many of these diets. (The book was published in 1963 by the Charles C. Thomas Co., Springfield, Illinois.) Several companies sell diet supplements which provide all the essential amino acids with just enough phenylalanine to meet the body's immediate needs.

Children who are put on proper diets within six months after birth can

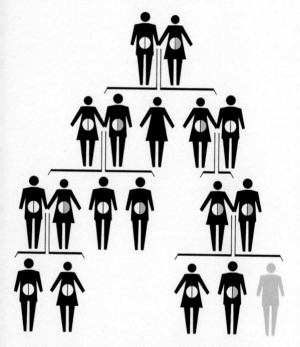

Each parent, in general, contributes a gene for each trait or characteristic a child possesses. The PKU condition occurs only when both parents give a child the gene that lacks the ability to produce an essential enzyme.

be expected to grow up with normal intelligence. If the diet is delayed beyond six months, some brain damage occurs and this is not reversible for more than 10 to 30 IQ points. After four years, all the diet can do is prevent additional mental retardation.

People who cannot metabolize phenylalanine also fail to develop normal skin and hair pigmentation. This shows how interrelated are so many body processes. Melanine, the pigment that colors the skin and hair, cannot be put together without the enzyme tyrosinase. This enzyme calls for tyrosine, which comes from taking apart phenylalanine; this can be done only when the liver produces the enzyme phenylalanine hydroxylase.

Born each year in the United States are an estimated 8 to 10 thousand babies whose genes do not permit the production of phenylalanine hydroxylase. Without the happy ending to this story, the lives of these people would be, at best, severely limited, and the expense of their care would be a severe drain on family resources. Among all the causes of mental retardation, no other produces such a high proportion of almost complete retardation.

Eliminating this kind of threat to so many babies can well be one of the proudest achievements of the men and women who match their wits and hands against the challenge of understanding molecules.

For good health, enough energy must be available to all tissues

The only way energy becomes available for all the things our bodies do is by the taking apart of certain kinds of molecules. Molecules of carbohydrates—the sugars and starches—are good energy sources. These molecules break apart and release energy when acted on by relatively mild collisions with enzymes and water molecules. Enzymes often have thousands of atoms per molecule. How they do what they do is difficult to understand but we know they break up food molecules very effectively. The enzyme in saliva that helps break starches down to sugars, for example, is so effective that we need chew a bite of bread only seconds before it tastes sweeter.

Within our bodies, enzymes allow carbohydrates to be "burned" by combining with oxygen at temperatures much lower than those at which sugar or starch burns outside living systems. In both cases, however, energy is released, and water and carbon dioxide are produced as byproducts or wastes.

Other kinds of molecules, especially fats, can also be taken apart to release energy. Some fats are solid like lard or tallow; others are liquids like corn oil, peanut oil, or olive oil. All fat molecules consist of carbon, hydrogen, and oxygen atoms, and can be "burned" to release energy if oxygen and the proper enzymes are available. As everyone who has tried to lose weight knows, fat is as likely to be stored in the body as it is to be burned to release energy. For many people, body fat doesn't disappear so long as we eat as much food as we want each day. To add to this problem, excess food of all kinds is likely to be changed to fat.

As pointed out in other stories, proteins provide building blocks that are needed to build cells and tissues. Proteins also serve as a source of energy. What happens to proteins in our bodies at any moment is a matter of balance. When adequate energy-releasing foods are available, proteins can be used for the growth and repair of cells and tissues. When other foods are in short supply, proteins must be taken apart to provide energy. Because protein molecules contain nitrogen, and sometimes sulfur, in addition to carbon, hydrogen, and oxygen atoms, the waste materials from "burning" proteins include other kinds of molecules besides water and carbon dioxide.

Good health requires protection against microorganisms

Disease sets in when certain microorganisms invade our bodies in nummers large enough to disturb the delicate balance among the many chemical reactions which are required for good health. One of the important roles of chemistry is to put together medicines that help our bodies restore balance and thus overcome disease.

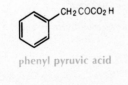

phenyl pyruvic acid

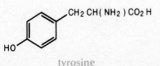

tyrosine

In the early days of medicine, cures and preventives of all kinds of disease were sought in the plant world. Books were published which described numerous roots, barks, stems, or leaves which allegedly cured every disease imaginable. Many generations of people have relied on "herb doctors" to provide the substances they needed to restore to body processes the rhythm and balance required for good health.

Some of these home remedies continued to be popular well into the 1900s. Many people can recall cough syrups, for example, that were made by boiling wild cherry bark, and adding vinegar, a bit of alum, a small onion, and a spoonful of honey. Goldenseal, ginseng, and May apples provided medicines many people believed in. Medicine for heart trouble was digitalis roots; lady slipper roots boiled in milk produced a kind of tranquilizer; and catnip tea was a standard cure for babies' upset stomachs.

The Middle Ages are remembered for their many elixirs, potions, and philters administered sometimes with and sometimes without magic incantations. Bloodletting has a long, long history and the practice continued well into the 1800s. In Paris, for example, leeches were used by the millions each year during the 1830s to bleed patients. It has always been difficult to keep medicine and magic, superstition and cure-alls from becoming entangled. So long as people have ills, real or imagined, there will be those who sell hope.

Chemistry attacks disease molecule against molecule

In past years, when home-made concoctions failed, people pretty much took for granted the aches and pains, suffering and death, "the ills that flesh is heir to." But this didn't keep a few men and women from dreaming of creating substances that would attack diseases head-on. These dreams really began to approach reality when Louis Pasteur proved that microorganisms rather than evil spirits or other mysterious forces caused many diseases.

It was the German chemist Paul Ehrlich who made one of the first giant steps toward proving that molecules could be put together in test tubes and flasks—molecules that would cure or prevent diseases as well as, or better than, the molecules that plants produce. In 1904, he found that compounds containing arsenic killed the tiny spirochetes which cause syphilis. He also proved that certain dyes would kill the trypanosomes, another kind of microorganism, that cause African sleeping sickness.

In 1935, G. Domagk, also a German chemist, found that the dye prontosil was effective against many kinds of streptococcal bacteria, the causes of serious "strep" infections. However, this dye did not kill the bacteria when they were grown in laboratory glassware. Investigation of this puzzling situation revealed that people who had been given prontosil excreted sulfanilamide. This was the molecule that actually got rid of the microorganisms.

Ehrlich's cure for syphilis, his "Magic Bullet," was widely dramatized, but the new sulfanilamide and the other "sulfa" drugs which followed were hailed as the miracle drugs of all time. People who remember the days before the late 1930s know why. "Strep" throats, infection in the mastoid bone, blood poisoning, all of these diseases were serious, even fatal threats.

There were additional reasons why the sulfa drugs created a major milestone in the relentless battle against disease. Chemists learned that the sulfanilamide molecule could be modified and each modification stood a good chance of being increasingly effective against different kinds and strains of bacteria. The research effort that was launched by the sulfa drugs provided the people and facilities that were needed to find out how bacteria actually cause disease.

Sulfonamides prevent bacteria from putting together a vitamin essential to them, folic acid. One of the raw materials for making folic acid is para-aminobenzoic acid. Structurally, this molecule is quite similar to a molecule of sulfanilamide. It is similar enough to slip into the spot where a para-aminobenzoic acid building block is supposed to go. It sticks there but won't allow the remaining steps required to make folic acid.

This realization led to the idea of "antimetabolites," drugs whose molecules are enough like a microbe's essential molecules to "get into the

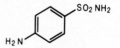

sulfanilamide

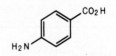

para-aminobenzoic acid

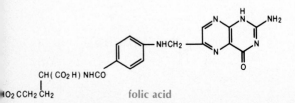

folic acid

24

act," but different enough to stop further action, especially the growth of more bacteria.

The story of penicillin

About the same time that the idea of antimetabolites was forming in the minds of men and women, people also realized that microbes compete with one another. One kind of bacteria or virus does things to the environment which keep other microbes out or kill those already there. The soil, for example, seems to be the part of our environment where many kinds of bacteria compete for survival, some more successfully than others. Much research has been based on collecting soil samples that might culture the growth of bacteriophages or whatever it was that caused one kind of microorganism to inhibit the growth of another.

The story of penicillin began when a stray mold drifted through a window of London's St. Mary's Hospital in 1929 and landed on an open petri dish in Alexander Fleming's laboratory. To many other scientists, this could have been nothing more than another contaminated bacteria culture. But Dr. Fleming noticed and chose to look into the reasons for no bacterial growth in the culture medium surrounding the colony of stray mold.

Some time later, Dr. Fleming attended a lecture where Gerhard Domagk told of his discovery of prontosil. After the lecture, Dr. Fleming said to a neighbor, "You know, I've got something much better than prontosil, but no one will listen to me. I can't get anyone interested in it. No chemist will extract it for me."

Not until 1940 were two Oxford men, Howard Florey and Ernest B. Chain, able to report that enough penicillin had been extracted to enable them to try it out on mice that had been given lethal doses of streptococci. Twenty-four of the 25 injected mice recovered.

Penicillin's story shifted across the Atlantic. In contrast to London, research and development facilities in the U.S. had not been destroyed by the bombs of World War II. Here a team of British and American scientists and engineers tackled the problems involved in producing commercial quantities of the antibiotic penicillin.

In March of 1942, the 33-year-old wife of a Yale faculty member received her first injection of penicillin. For weeks, her health had declined as a result of infection after a miscarriage. Despite three operations, six transfusions, and the drugs that were available at the time, her streptococcic infection threatened to claim another victim. Penicillin was injected every four hours. Her temperature dropped within 12 hours from 105.5° to 98°F. From this point, her recovery was uneventful and she was discharged from the hospital.

The almost miraculous result from this one clinical test explained why people were eager to solve the problems of commercial production of penicillin. These problems, however, were difficult. The *Penicillium* which produce penicillin required carefully controlled environments. Large quantities of culture medium were needed to yield tiny quantities of the drug. At first, only one part of penicillin could be harvested from one million parts of the fermentation broth.

The penicillin molecules were being created deep within the world of microorganisms—the world where taking molecules apart is much more the order of the day than putting complex molecules together. This is where bacteria thrive, where complex molecules provide the energy and building blocks which enable simple organisms to duplicate themselves amazingly fast. Unless all competing microorganisms were kept out of the fermentation tanks, penicillin molecules could not accumulate.

War creates a whole new dimension of drives and motives. From one point of view, only the randomness of the draft lottery determines who lives and who dies. Those who live must compensate for those who do not. Motivation such as this may have speeded the effort to produce commercial quantities of penicillin. Actually, good luck favored the knowledge and skill of the team of chemists, biologists, engineers, and physicists who worked on the mass production of penicillin.

In a Peoria, Illinois food market a government research worker found a moldy cantaloupe on which was growing a new strain of *Penicillium*

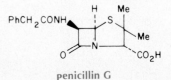

penicillin G

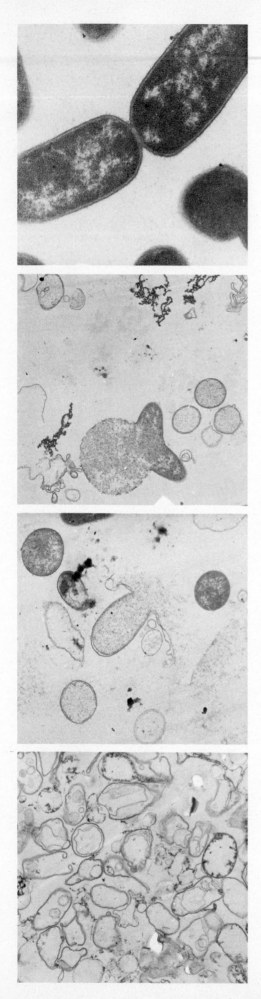

Here we see how penicillin fights *Escherichia coli,* **one of the bacteria which cause many infections. The infections set in when the bacteria multiply. To multiply, the cell wall of a bacterium cell must be taken apart and new cell walls put together around two new daughter cells. This is where penicillin comes in. The penicillin molecules interfere with the enzyme which controls the cross linkage of the molecules which form cell walls. Without cell walls, the bacteria cannot multiply and the cause of the infection is removed.**

chrysogenum. This strain thrived in deep tanks of culture medium and yielded 200 times more penicillin than Fleming's original mold. The new mold also flourished on a medium consisting for the most part of water in which corn had been soaked, a medium that was available in large quantities. Production jumped from fractions of a gram to tons. It was used in Normandy on D-day and every day thereafter throughout World War II. By the end of the war, the pharmaceutical industry was producing enough penicillin to treat seven million patients.

World War II ended. We cannot say how many lives were saved by penicillin. But in the world of microorganisms, the struggle to survive goes on and on. In this world where relatively simple organisms exist, where large molecules are being taken apart, there are many opportunities for new kinds of molecules to be created. Or, in the language of biologists, mutations can occur and new strains of microorganisms appear —strains resistant to penicillin.

There was another problem which threatened the usefulness of penicillin. Although it was generally nontoxic, some patients developed sensitivity to this antibiotic. To solve this problem, a precedent was set by the sulfa drugs. By modification of the culture medium in which *Penicillium* grows, a substance which can be used as the building block to produce different varieties of penicillin molecules can be made to accumulate. Each modified molecule holds promise of being uniquely effective against new strains of microorganisms as well as being less toxic to sensitive people. In 1958, Dr. John C. Sheehan earned a patent on this process for creating modifications of the penicillin molecule. Production of the basic building block, however, still requires the assistance of some of nature's most simple organisms.

Penicillin was the first of the antibiotics harvested from molds. The stories of other antibiotics deserve to be told also. In any case, the lives of thousands of men and women, boys and girls, will always be a memorial to the discovery and production of this amazingly effective family of drugs.

Some kinds of molecules provide continuing immunity to disease

Measles, mumps, chickenpox, and a few other diseases can be suffered only once in a lifetime. Furthermore, diseases often run their course and good health returns even though no medicines or drugs are used to cure the diseases.

These facts pose the challenge to find out exactly what kinds of molecules provide immunity or natural cures, how these molecules are put together, and how they can be produced. If this challenge can be met, then these molecules could be used to back up the body's natural disease-fighting mechanisms.

Immunity involves antigens and antibodies. Antigens are molecules which do not fit into the chemistry of healthy bodies. They can be brought into the body by disease-producing organisms through the body openings, and, as in the case of such a notorious antigen as poison ivy, through the skin. Antibodies are molecules which the body produces when antigens enter the body and which counteract the effects of the antigens and thereby bring about cures or immunity.

The molecules which cause immunity are not only very complex but also extremely low in concentration. Immunoglobulin, an antibody in plasma, for example, is put together from four chains of amino acid groups with several hundred amino acid building blocks in each chain. Remember that a single amino acid unit out of its proper position could cause the whole molecule to be ineffective.

The first time an antigen invades the body, a small amount of antibody is produced by the thymus, spleen, lymph nodes, or by two families of cells, the macrophages and lymphocytes. If the antigen invades a second time, much more antibody is produced. The cells producing the antibody disappear, however, and the antibody level falls off. But when another invasion occurs, high antibody levels reappear promptly. Much of the mystery of the antigen-antibody interaction involves how the body remembers previous antigen invasions, what kinds of particles do the remembering, and where they are between invasions.

One theory proposes that the antibody molecule becomes a part of lymphocytes, perhaps attached to the surface. Antigen molecules could become attached as they float by in the blood or lymph. This could trigger rapid production of additional antibody molecules.

Antigen-antibody action is prominent in medical research today, and for several reasons. There is the haunting possibility that science could find an antibody that would cause the body to reject cancer cells or tumors. A second reason involves the replacement of defective hearts, kidneys, lungs, livers, skin, or other organs. It is ironical that the body's own antigen-antibody reaction is the most serious threat to the success of organ transplants. It is almost impossible to match the tissues of the donor and the recipient closely enough to insure that the recipient's antibody system doesn't treat the new transplant as an antigen and destroy it.

A mother's body tolerates the cells and tissues of her embryo during pregnancy. Because the father's genes also influence the proteins that are put together in the embryo, by no means are the embryo's molecules perfect matches of the mother's. This suggests that there are molecules which can override antigen-antibody action. Searching for these so-called immunosuppressant substances is an exciting part of today's medical research.

The story of immunity against polio

Without doubt, this story has the happiest ending of all of the stories about people matching their wits and hands against disease. To begin, however, here is one of the thousands of sad case records of polio victims which were written each year before the late 1960s.

Case 4. Male student, age 19. Days 1-3: general malaise; felt progressively more "fed-up." Day 4: went for a cross-country run (four miles) to try to "work off" his malaise; he did not run easily, getting home with a struggle. Day 5: attended lectures; no marked symptoms. Day 6: fever and headache; walked with difficulty 200 yards to tell a relative he was ill; returned to bed. Day 7: during the night he felt ill and weak when he got up for a drink; in the morning he was helped to the lavatory, but collapsed on the floor. This was the last time he walked. Paralysis was almost complete and permanent in both lower limbs and lower trunk.

Here ended one case history insofar as hospital records were concerned. Polio's full story, however, continued through untold numbers of lives lost outright or spent on crutches, in wheelchairs, or in iron lungs. Between 1915 and 1955, 528,851 cases were reported to health officials. As many as 40,000 of these were fatal. Twice this many polio victims needed crutches, braces, or wheelchairs after recovering. More than half of polio's victims were less than 10 years old. This one disease accounted for more than one-fifth of all skeletal and muscular deformities.

Early in the 1900s, poliomyelitis was proved to be caused by a virus, one of those particles which seem to link the living and non-living worlds, living because they multiply when the environment permits and exist for long periods of time in apparently non-living condition when the environment demands. From many points of view, viruses are simply highly complex molecules.

When people, especially very young people, are exposed to poliovirus,

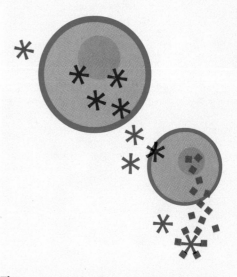

The response of cells to invasions by virus or other "enemy" particles is some of the body's most miraculous and mysterious chemistry. Natural immunity is triggered by virus (red shapes) entering healthy cells. This stimulates the production of antibodies (green) which help to destroy the virus.

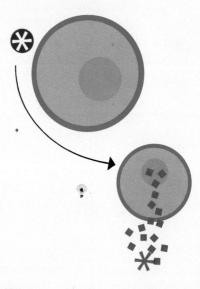

Artificial immunity is triggered by introducing dead particles (purple) which retain the ability to stimulate antibody production. These antibodies, in turn, head off future attacks by live virus.

Years ago, many babies were exposed to mild polio infections before they lost their mothers' immunity providing antibodies.
With this help they were able to fight off the disease and retain immunity.
However, with today's higher living standards, babies are no longer exposed to the disease and thus do not build up early immunity.
Ironically, shielding our children from polio may have added urgency to finding a preventive.

the virus particles settle in the brain or spinal cord. Long nerve fibers extend from the nerve cells in the brain and spinal cord to the muscles. One nerve may control as many as 400 separate muscle fibers and thousands of fibers are bundled together to make a single muscle. Poliovirus attacks nerve cells and, if it is present in large enough quantities, the nerve cells are destroyed beyond the point of repair or recovery.

By the time they are adults, most people have become immune to polio. Looking back at the massive efforts of men and women to conquer the disease, we find that this was both an advantage and a handicap. It was an advantage to know that immunization was possible. It was a handicap never to be sure who had and who hadn't acquired natural immunity, or how and when immunity was acquired.

The beginning of polio's story is hidden in the backlog of knowledge and skills that has accumulated through many, many years of science and medicine. A good example of this is the work of Alexis Carrel. In 1902, he took up the work of Ross G. Harrison and learned how to grow animal tissues in glass containers—tissues which heretofore could be grown only in the more complex environment of an animal's entire body. Dr. Carrel received the Nobel prize in 1912 for this achievement.

H. B. and M. C. Maitland in 1928 made an almost equally giant step when they learned how to grow viruses in glassware. They grew vaccinia virus in finely divided hen's kidney tissue and serum. Improvements in their process, which were developed in the late 1930s, involved adding something to the culture medium—the sulfonamides, for example—to kill competing bacteria. Progress also increased when it was found that hen's eggs provide an almost ideal container and culture medium for growing microorganisms. But no one could grow poliovirus.

A breakthrough came in 1936. Doctors A. B. Sabin and P. K. Olitsky, apparently taking a lesson from the poliovirus, prepared culture media containing tissues taken from 3-4 month old human embryos. Each medium was prepared with a single kind of tissue. The medium that contained brain and spinal cord tissue grew the poliovirus.

By this time it was known that there were at least three strains of poliovirus. These strains were different enough to make it all the more difficult to pin down the virus. At the same time, because most polio victims recovered with no permanent paralysis and because millions of people acquired immunity without realizing they had been exposed to the disease, it was easy to believe that there might be a strain of the poliovirus that would produce immunity without risk of paralysis or death.

In 1949, J. F. Enders, T. H. Weller, and F. C. Robbins used Dr. George Gey's method of gently rolling over and over the flasks that contained the culture medium in which poliovirus was to be grown. This approach grew the virus in media that were easier to obtain than those containing human embryo tissues. They also found ways to identify the virus by using microscopes rather than by injecting material suspected to contain poliovirus into the brain of a monkey. This earlier method of identifying the virus required the sacrifice of more than one monkey for each test.

By 1951, 13 strains of poliovirus had been grown in cultures of living cells, particularly embryonic intestinal tissues. It is a big advantage to be able to grow poliovirus in laboratory glassware rather than having to study the virus and its effects in polio patients. In glassware, it was possible to dig out the biochemical factors and to seek the actual molecules involved in the growth and multiplication of the virus. People could foresee being able to bring together the kinds of molecules that would enable children to be immunized against polio.

Actually, the idea of an immunizing vaccine for polio goes back to the 1920s. At that time a vaccine was prepared by killing the virus with heat before injecting it into monkeys. Other investigators tried to produce a vaccine by inactivating the virus with carbolic acid, formaldehyde, and other substances known to be harmful to microorganisms.

In 1935, a polio investigator created a vaccine consisting of a 10% emulsion of polio-infected monkey spinal cord tissue that had been treated with formaldehyde. He had enough faith in this vaccine to test it on 3000 children. When some of these children came down with polio, he knew his faith was unjustified. He was criticized so severely by his colleagues for these disastrous results that he was hard put to find a place to continue his work.

At this same time, another investigator created a vaccine in which he had great faith. His vaccine was also based on weakening the virus enough to prevent it from causing the disease but not totally killing the virus. He tried the vaccine on himself and his two children. When he tried it on 1000 other children, however, 10 of these children came down with polio. This was 500 times greater than the rate at which children had polio even during the most epidemic years.

As a result of these premature vaccines, a wave of revulsion against human vaccination trials of polio vaccine swept through the scientific and medical professions, affecting the behavior of research people for many years. Circumstances such as these are most thought-provoking. When we realize the fear in the minds of people who knew the horrors of polio and its ever-present threat to their children, it is easy to understand the sense of urgency that would drive people to take chances on possible cures or preventives. In contrast, it is equally easy to be critical of undue caution when we realize that each year a potentially successful vaccine was kept off the market another 35,000 or more became polio victims (in the U.S. alone).

World War II may have affected the timing of the ultimate production of an effective and safe polio vaccine. Many of the people who had worked on polio found themselves with totally different assignments during the war—different at least in the kinds of disease being studied.

Dr. Albert B. Sabin is a good example. He managed immunization programs involving other diseases in Okinawa, Japan, China, and other parts of the world. These assignments not only gave him valuable experience but also allowed him to rethink and rearrange his ideas about polio.

The end of World War II was marked by some very bad polio years. In 1946, for example, 25,191 cases were reported. This was more than twice the average of the preceding five years. In 1947 there were 10,734 cases, but the number went back up to 27,902 cases in 1948, 42,202 cases in 1949, dropped to 33,202 in 1950, and then stayed around 38,000 cases per year for 1951-1955.

Following World War II, another idea shows through the research being done on polio. Strains of poliovirus seemed to lose their potency after being cultivated through several generations in the laboratory. Cultures of highly virulent strains were much less likely to cause paralysis after the cultures had been transferred repeatedly to new culture bottles.

In 1951, H. Kaprowski took the bold step of feeding 20 volunteers a strain of poliovirus that had been modified by producing itself in laboratory rats. None of these volunteers came down with polio and their body fluids showed a marked increase in the antibodies which cause immunity. Dr. Kaprowski continued research in this direction.

By 1954, however, Dr. Jonas Salk was further along on reviving the idea of inactivating the poliovirus with formaldehyde. He gained an assist from Dr. Jules Freund's 1942 discovery that the effectiveness of vaccines can be increased by whipping the vaccine into creamy emulsion with specially purified mineral oil.

In 1952, Dr. Salk prepared and tested a vaccine containing three strains of formaldehyde-modified poliovirus in mineral oil. After testing the vaccine on more than 100 children and adults, he announced the successful results on March 28, 1953. This vaccine quickly captured the confidence not only of Dr. Salk's colleagues but also of the public. Mass production of the vaccine moved ahead promptly, and by 1954 it was being given to millions of children and adults. By the 1960s, the number of polio cases being reported had dropped to the hundreds, and people began thinking of the day when polio would be no more.

Not often are two solutions to the same problem found at about the same time. It is quite remarkable that Drs. Kaprowski and Sabin were completing their work on a different vaccine at about the time Dr. Salk announced his success. The second vaccine was based on producing harmless strains of the poliovirus—harmless because they had been carried through enough generations to have lost their ability to cause polio but were still able to cause immunity.

This vaccine was given to many people in Ireland in 1956 and to 250,000 people in Africa in 1957. Twenty large-scale trials involving millions of people were carried out in 1958-59. In effect, by 1960 the world now had two vaccines, either of which could conquer polio.

1902	Tissues grown in lab containers
1928	Virus grown in glassware
1936	Discovery that virus grows in brain tissue
1949	Virus indentified by microscope
	Salk and Sabin vaccine administered to millions

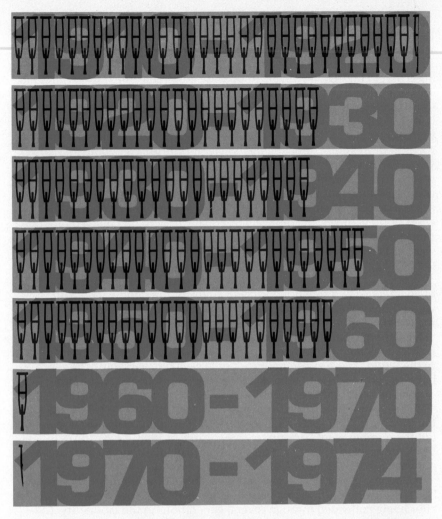

 = 500 deaths, USA

Between the lines of these dramatic data is the lifework of great scientists.

This turned out to be good. In 1955, certain batches of the Salk vaccine in which the poliovirus had not been totally killed by the formaldehyde were put on the market. Several people developed polio from the vaccine. This tragedy points out much of the stress and strain that goes with mass production of drugs and medicines.

It is to be expected that people who are devoted to saving lives and alleviating suffering are particularly sensitive to accidents that cause death and suffering. The author happened to visit the president of the company that produced one of the fatal batches of vaccine. He had just returned from a brief vacation and seemed to be recovering from the shock caused by the accident. While a group of teachers were being conducted through his organization, it became clear that any one of hundreds of his employees could have been responsible for the error that produced the fatal batch of vaccine. But at no time did he even suggest that anyone but himself was responsible for the tragedy. It was equally evident that he was doing everything he could to make sure the error would never happen again.

The same evening there appeared on the cover of a magazine a photo of this man and the announcement that the lead article dealt with his "criminal record." This magazine article meant little to the millions of people, especially the parents of young children, who knew that the threat of death, crutches, braces, and wheelchairs was over.

But the memory and impact of that sort of publicity lingers on. Scientists are taught valuable lessons, and so must be the public. No new drug or medicine can be introduced without a degree of risk. No matter how great are the precautions, accidents will happen. No matter how many checks we have on premature action, unfortunate situations will occur.

30

The important thing is that research will continue. The motivations and support of the men and women who must do this research cannot be frittered away or neutralized by irresponsible publicity.

The polio story ends on a truly happy note. It is a success story at a time when all too many scientists are all too aware of problems for which solutions have not yet been found. True, the men and women whose careers are devoted to taking things apart and putting things together have played only a supporting role in the polio story but it is a proud one.

Every body needs vitamins

One hundred years ago it was observed that laboratory animals do not thrive on a diet consisting of all of the substances believed to be present in milk. This was quite puzzling. Milk is the only food newborn animals receive, and they usually thrive on it. Almost as puzzling was the fact that eating fresh limes kept British sailors (forever after nicknamed "limeys") from getting scurvy when long voyages made other fresh fruits and vegetables unavailable.

Observations such as these hinted increasingly during the early 1900s that good health required something in addition to carbohydrates, fats, proteins, and minerals. In 1912 Casimir Funk named the missing "something" vitamines, a name later shortened to vitamins. A search for these substances became worldwide. The dramatic stories of the discoveries of the individual vitamins have been told many times—and rightly so. These stories illustrate very well what chemistry is, what chemistry does, and what the results have been. The results of vitamin research were that people have practically forgotten about such diseases as scurvy, beriberi, rickets, sprue, xerophthalmia, pernicious anemia, and pellagra.

Glimpses of the story of vitamin K illustrate how many of the vitamins were discovered. The vitamin K story began when physicians became puzzled by the bleeding that threatened the lives of people with liver disease, especially if the bile duct was blocked as it sometimes is in newborn infants. In the 1900s, George Whipple showed that such bleeding seemed to be related to inadequate prothrombin, a substance in plasma that helps blood to clot.

About this same time, Henrik Dam at the University of Copenhagen was studying the effects of removing sterols, particularly cholesterol, from the diets of chickens. He was surprised to see the chickens develop bleeding under the skin. Later it was found that this bleeding could be prevented by feeding fresh cabbage, and that the bleeding was caused by inadequate thrombin. Acting on a hunch suggested by these observations, Dam set out to discover a new vitamin he believed existed in green leaves and liver. He was successful and called the new substance, vitamin K.

Another part of the vitamin K story happened in 1933 when a farmer came to the University of Wisconsin with a dead heifer, a milk can of blood that wouldn't coagulate, and a bale of spoiled sweetclover hay. He remembered a disease that swept North Dakota in the early 1920s that was caused by feeding improperly cured sweetclover to cows. By chance, the farmer went to R. P. Link who was interested in this disease. Dr. Link went to work on the farmer's problem, and by 1939 he had isolated the specific molecule in the spoiled hay that had caused the trouble. Equally important was his discovery that the action of this molecule, dicoumarol, could be overcome by vitamin K.

In keeping with the way chemists work, it was realized that dicoumarol might help people whose blood shows too great a tendency to coagulate. Tests showed that dicoumarol could be used to control the effects of prothrombin and could be teamed with vitamin K to bring blood clotting pretty much under control.

In a project only remotely related to bleeding caused by a vitamin deficiency, Dr. Link decided that dicoumarol might be a good rat poison. It is colorless and odorless, and large doses should cause fatal bleeding in rats. But dicoumarol proved to be too weak an anticoagulant. Consequently, chemists turned to a similar molecule, warfarin. Because warfarin is quite safe for people, millions of pounds of this compound are used each year in the battle against rats and mice.

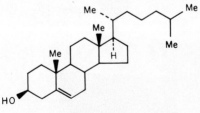

cholesterol

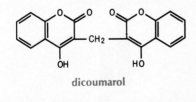

dicoumarol

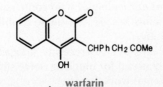

warfarin

There are times when we need to feel no pain

And not just because pain hurts. Surgeons cannot repair damage or correct defects in our bodies unless we can be put to sleep and our muscles relaxed. From many points of view, the development of safe and effective anesthetics could well be the greatest single contribution of chemistry to the well-being of people.

The effects of anesthesia are also among the most mysterious of the effects of chemistry on our bodies. But, as someone has put it, how can we expect to understand how chemicals put us to sleep when we do not yet understand consciousness?

People have known for a long time that there are substances which affect our perception of stimuli. Opium is probably the best known, but all people, no matter how primitive they may be, know of some kind of plant or animal product that takes the edge off pain. Alcohol has been used widely as a sort of anesthetic.

Sir Humphry Davy in 1798 noticed that nitrous oxide, a colorless gas with a slight sweet taste and odor, seemed to affect a person's keeping in touch with things. Because of its strange effects it was called laughing gas. It was not until 1844, however, that anyone used laughing gas as an anesthetic.

Crawford W. Long discovered the sleep-producing effects of ether in 1842, and seven years later Dr. W. T. C. Morton, a dentist, encouraged Dr. J. C. Warren to give ether to a surgery patient. Chloroform was also used as an anesthetic about this same time, although it had been known as a compound as early as 1831.

Quite probably, only sheer curiosity revealed the pain-killing effects of nitrous oxide, ether, or chloroform. These three substances aren't very similar, although each is less soluble in water than in fats and oils. Because these substances are highly volatile, they can be breathed into the body and then be taken up by the blood and carried quickly to all body tissues. Nerve fibers are covered with myelin, a substance that contains molecules of fat. This is where the high solubility of anesthetics comes in. In effect, the nitrous oxide, ether, or chloroform molecules are pulled from the blood and absorbed in the tissues which surround the nerves. Not until the absorbed substances are gradually diffused from the nerve tissue will the nerves be able to communicate pain.

Being put to sleep by an anesthetic shows the delicate balance among substances in our cells and tissues. Body processes can be influenced and the amount of influence controlled by the amount of anesthetic used. An anesthetic that is especially soluble in fats can affect nerve tissue without upsetting other body functions. This is exactly what is needed to abolish pain and relax muscles.

The development of an anesthetic that relaxes muscles also shows how chemists have contributed to the success of modern surgery. This story began, strangely, in the jungles of South America. As early as the 1500s, Spanish explorers found the Indians of Peru using arrows which carried an amazingly deadly poison. The effects of the poison were so spectacular that the Indians were coaxed to reveal the recipe. Actually, there were several recipes for the poison but each recipe relied on the bark from the roots or stems of four or five different plants. The bark was scraped from the plant, crushed and shredded, then boiled for several hours. The resulting soup was strained through a bed of leaves and then boiled down until it was a black, sticky tar. The poison was stored in hollow canes until needed.

To prove that each new batch was effective, a pin was dipped in the poison and used to prick a lizard's toe. If the lizard was dead in less than ten minutes, the batch was good. When stories of this deadly poison were first told to Europeans, it was natural that the stories were given at least a dash of superstition. Moreover, it could have been that sometimes the Indians threw in such ingredients as the teeth and livers of snakes, red pepper, and similar additives.

Eventually, the active substance in many of the recipes for arrow and spear poisons was proved to be curare. By 1930, pharmaceutical chemists put this molecule on the market in a form that could be administered in precisely controlled quantities.

The differences between using curare to kill game rather than to cure

toothache!

neuron

axon

dendrite filaments

anesthetics

Molecules which enable dentists and surgeons to control our feeling of pain come close to being chemistry's greatest single contribution to humanity.

Novacaine, or more recently, lidocaine is especially useful for blocking nerve stimuli during dental work.

today's illnesses reveal much of what has happened between the 1500s and the 1900s. Actually, it is difficult to associate the white-coat environment of a modern pharmaceutical laboratory with the superstition-filled environment which first created curare.

There are, however, threads of similarity between the two widely different environments. In both cases, the goal is to extract from a mixture of raw materials the specific kind of molecule that is needed to do a specific job. When this molecule is put into the fluids which surround nerve cells and fibers, the action of the nerves—especially those which control muscles—is abolished.

The characteristics of the curare molecule are built into the molecule because its atoms are the kinds they are and because of the way they are arranged. Today's chemists know what the molecule looks like. They also know which kinds of molecules in our cells and tissues the curare molecule interacts with. This information enables physicians to determine exactly the proper quantity of the drug to use for each operation. Neither the white-coated chemist nor the Peruvian Indian, however, knows why the curare molecule has the characteristics it has. But this doesn't keep the molecule from relaxing a surgery patient's muscles or numbing the muscles of an animal that is needed for the Indian's supper.

Today's chemists can put together the curare molecule from raw materials that are much more easily obtained than the jungle plants used by the Indians. By a slight remodeling of the molecule, unwanted side effects are minimized. In fact, when the molecule-to-molecule effects of curare were clarified, it became possible to put together a new substance, succinylcholine chloride, which does what curare does but in a way that makes it a much safer drug.

It was proved that curare blocks the activity of acetylcholine. This is the molecule which transmits nerve stimuli to muscles. Apparently the only way a nerve stimulus can jump from a nerve ending to a muscle fiber is by the liberation of acetylcholine at the junction between nerve and muscle. When curare molecules are present in the fluids surrounding these junctions, nerve impulses cannot pass.

Curare acts first on the muscles which are controlled by cranial nerves,

South American Indians concoct the deadly poison, curare, by shredding the bark from the roots of certain plants, adding water, then boiling the mixture until it is reduced to a sticky tar. Darts are then dipped in this poison and used to kill small game, such as monkeys, for food.

Much of the progress of chemistry consists of understanding and improving our use of nature's molecules to solve problems.

The curare molecule has been isolated and modified for use as a muscle relaxer.

33

then on muscles controlled by nerves of the trunk, the arms and the legs, and, finally, the curare acts on the muscles which control breathing. By controlling carefully the amount of the drug injected into the patient, the surgeon can determine which group of muscles will be relaxed and for how long.

Molecules and mental health

Good health depends on an intricate assembly of fantastically complex chemical reactions and good mental health requires a very delicate balance among all these reactions. From the beginning of recorded history people have suffered mental illness and they still do. In the United States alone, more than 30 thousand hospital beds are needed to care for schizophrenics, people who have lost the ability to cope with situations and have withdrawn into a world of fantasy. More patients throughout the world are probably being treated with Valium than any other drug. And this is only one of the drugs used to reduce tension and anxiety.

The history of people's reactions to mental disease does not make a pretty story. Far too much of this history features evil spirits or demons that had to be exorcised from bodies of which they had taken possession. When prayer or magic didn't work, or purgatives wouldn't drive out the evil spirits, the "possessed" were starved, flogged, and even burned or stoned to death in order to make their bodies unpleasant dwelling places for evil spirits.

Dark, filthy cells in "insane asylums" where mental patients were confined, although supposedly a scene from the Middle Ages, are too much like the conditions which existed well into the 1900s in mental institutions. Philip Handler suggested in *Biology and the Future of Man*, published by the Oxford Press in 1970, that 90% of what was known about the biology of human behavior had been learned since World War II. He added that, if the picture seemed incomplete and less than satisfying, there was hope in knowing that neurophysiology and neurochemistry are attracting some of the brightest and most fertile minds.

And rightly so. Study of the mind is the ultimate in difficulty. Normal and abnormal behavior are separated by only dimly seen barriers. It is almost impossible to determine whether the behavior of a person at any moment is due to ongoing chemistry or is caused by memory. And unless the patient's mind has undergone massive intervention, there is always the haunting possibility that the patient selects the behavior the investigator is permitted to observe.

Despite all of these difficulties, glimpses of success are beginning to outweigh sheer frustration in mental health research. This shows through in a student's paper reporting her reading about schizophrenia. The paper follows with only a little editing:

Schizophrenia is a mental disorder in which the patient loses the ability to make and hold relationships and then withdraws into a world of fantasy. For a long time, it was believed that although the disease interferes with normal functioning, it does not have its origins in physiological causes.

Early ideas concerning schizophrenia associated the disorder with: traumatic experiences when young, heredity, or arteriosclerotic changes. Little came from these ideas.

Then came an influx of new ideas. It was found that a high percentage of schizophrenics are diabetics, also that 10% of schizophrenics respond slightly to thyroid hormone therapy. By the 1950s, it appeared that the disorder could be organic but it was not until the 1970s that there was some real proof.

Because they set the ball rolling for new advancements, the work of C. D. Wise and L. Stein seems to be the most important. They worked on the unproved theory that the noradrenergic pathways in the brains of schizophrenics are damaged. If these pathways are destroyed then schizophrenics are the way they are because their brains receive distorted "goal-oriented" messages.

Wise and Stein started out with this shaky idea but it is not where they ended. First they found that dopamine betahydroxylase (DBH) was in lower concentrations in schizophrenics than in normal people. There was 11 to 23% deficiency in some parts of the brain and as high as 50% in others. This was accompanied by low concentrations of noradrenaline, suggesting a disturbance of the noradrenergic neurons.

Another group of investigators found low levels of monamine oxidase activity

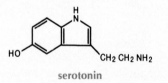

dopamine

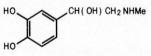

serotonin

adrenaline

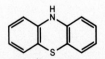

phenothiazine

34

A healthy mind perceives things as they are.

The diseased mind distorts the images it receives.

An uneducated mind cannot add meaning to what the eyes see.

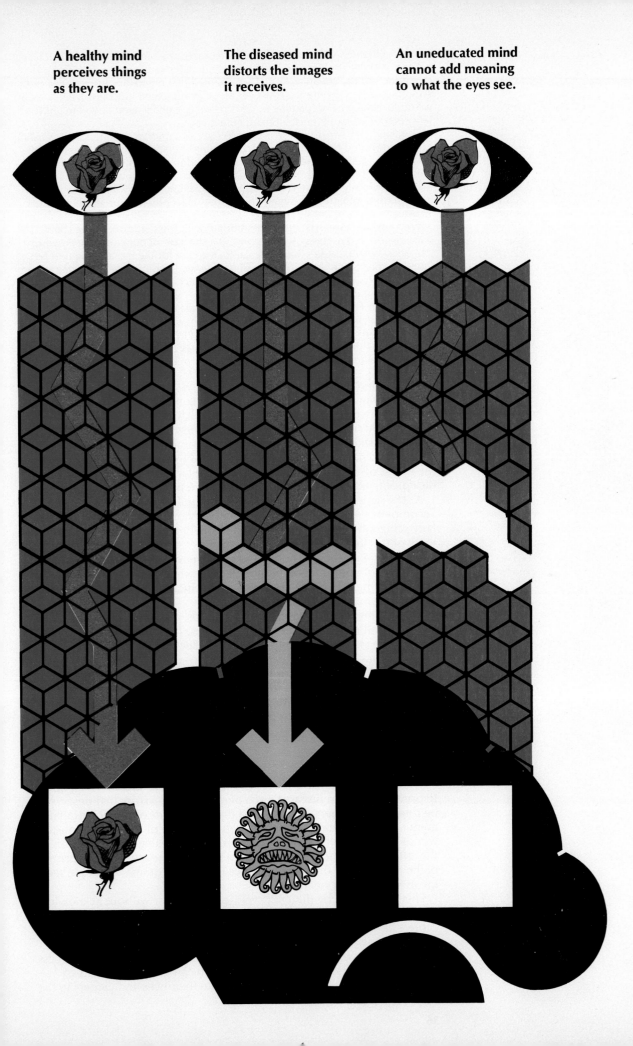

in schizophrenics. Other related studies reported the presence of abnormal levels of serotonin metabolism and that certain stimulants seemed to precipitate psychosis in "normal" people. Also, it was learned that phenathiazine reduced the symptoms of schizophrenia. However, dopamine seemed most involved. Now if we understand what dopamine is, perhaps we can understand how it might lead to a cure for schizophrenia. Sit tight, here goes.

Dopamine beta-hydroxylase is used to synthesize noradrenaline. Noradrenaline can be converted to adrenaline. Noradrenaline and adrenaline are stored in the nerve endings. Dopamine and trypsin are stored in the cytoplasm. All four of these kinds of molecules must be present to have nerve impulses pass from cell to cell. Obviously, if there is a shortage of dopamine, there will be irregularities in the passing of nerve messages. This is why schizophrenics are ill; they get incorrect messages.

We know for sure that drugs can interfere with the synthesis, storage, and transmission of messages by nerve cells. If a shortage of dopamine beta-hydroxylase does, in fact, cause schizophrenia, then a cure for the disease ought to be found in controlling the availability of dopamine. Phenothiazine provides an indirect approach. This drug blocks the nerve receptors which control the production of dopamine.

From what I see of this whole topic, there must be something missing in the DNA of the schizophrenics and the messages they get from their brains aren't the ones that were intended. There is much confusion in the research but some facts seem to be falling in place. More and more, however, people seem to agree that it is chemical imbalance that is destroying their minds.

This student's paper reflects the difficulties associated with efforts to describe precisely the conditions which accompany mental disease and create or discover drugs and medicines which prevent or cure mental disorders. Although she found "much confusion in the research," she was impressed by the urgency to attack this problem. Similarly, she was encouraged to see that reasonably clear-cut hypotheses are shaping up in the minds of the men and women who are tackling the exceedingly complex problem of mental disorder and disfunction.

At the same time, many false hopes and subsequent disappointments accompany efforts to trace the biochemistry of mental disorders, particularly schizophrenia. As Dr. Solomon H. Snyder points out in a 1974 summary of this field of research, " 'discoveries' of the biochemical abnormality in one or another body fluid . . . have relentlessly been followed by failures of confirmation in other laboratories." For sure, the reduced amounts of beta-hydroxylase in schizophrenics is an enticing albeit a tentative hypothesis.

Today, only heart disease, cancer, arthritis, and allergies cause more suffering and death than do mental diseases. This fact has had much to do with the launching of vigorous and productive efforts to find or create new drugs for mental disorders and all forms of disturbed mental behavior. During the 1950s, for example, chlorpromazine, a component of one type of anesthetic, was found to function as a tranquilizer. This discovery triggered efforts to put together similar molecules that might cure mental diseases. Nearly 3000 new molecules were created and tested.

A Look Back

Thousands of men and women have exhausted their lives seeking ways to prevent and cure diseases. Each of their lives is a story that deserves to be told. Insulin, phenylalanine, and the other kinds of molecules that are mentioned in this chapter are only representative of the many molecules whose roles in health and disease have been revealed. The sulfa drugs, the antibiotics, polio vaccine, vitamin K, dopamine—each is an example of the many, many dramatic episodes in the total drama of people seeking to improve our health.

Finally, it would be difficult to close this chapter without acknowledging the efforts of the highly dedicated men and women whose lives and careers have been spent looking in vain for the causes, cures, and preventives of diseases which continue to threaten our health and wellbeing. These people may never have known the satisfactions of success, but their work has created pathways which can guide the efforts of younger men and women who will devote their lives and careers to the chemistry of staying well.

3 Supporting Us: The Chemistry of Farming

The cereal grains, especially wheat, rice, and corn, produce both "animal food" and "people food." Chickens, turkeys, and other poultry convert animal food to people protein foods most efficiently, that is, with the least loss of the energy originally stored in the cereal grains.

The theme of this book is what chemistry is, what chemists do, and what the results have been. This chapter is concerned with what farming is, what farmers do, and what the results are. We are especially concerned, of course, with how chemistry helps farmers do their work.

Farmers raise and harvest the plants, and take care of the animals which produce our food and a good share of our textile fibers. In addition to supplying foods in seemingly unlimited quantities, farms and forests provide the raw materials for making lumber, leather, paper, clothing, and several other nonfood items.

Working closely with farmers are the people who process, preserve, package, store, and distribute our foods. The food industry provides jobs and careers for one-seventh of the working population. Many of the foods we like best are very perishable. Fruits, vegetables, meat, eggs, and dairy products don't keep unless refrigerated or preserved. If farmers did not work closely with the industries which take care of these products, we would not have year-round supplies of many of our most popular foods.

We almost take for granted supermarket shelves piled high with eye-catching packages, cans, and bottles of prepared foods. We assume that refrigerated bins will always be filled with fresh and frozen fruits and vegetables. We expect to find trays loaded with meat, seafoods, eggs, and dairy products. These are all familiar scenes.

Less familiar are the fields and orchards, feedlots and poultry houses, dairy barns and processing plants from which come all our supermarket food. For every 200 shoppers who reach for a carton of milk, for example, a farmer somewhere is feeding 10 cows, milking them twice each day, and protecting them from diseases and insect pests. Furthermore, these cows must take "maternity leave" every year or so if milk production is to continue.

In a recent year, the 11 million cows on the farms in the United States

38

produced more than 115 billion lbs. of milk. The average was 10,125 lbs. of milk per cow. An additional 819 million lbs. of milk and dairy products was brought in from other countries.

Farmers are in the energy business

Farmers use green plants to collect and store the sun's energy. The energy is neatly packaged in the products farmers market. This is the only source of energy we need to keep us alive. Included, of course, are the "farmers of the sea" and their harvest of fish and other seafoods.

The total relationship of farming and food processing to our energy resources is becoming increasingly complex. Using eggs as an example: in 1910 farmers spent less than one calorie of fossil fuel energy for each calorie of egg food energy they produced. Today each calorie of food energy produced costs five fossil fuel calories. In the U.S., food growing consumed four times as much energy in 1970 as in 1940. During this same time span, the energy used for processing foods tripled. The grand total of energy used in 1970 for producing, processing, refrigerating, and cooking our food was 2172 x 10^{12} kilocalories. The comparable figure in 1940 was 685.5 x 10^{12} calories.

The future for the world's people depends more and more on how efficiently farmers and the food industry capture and deliver the sun's energy in the kinds of food we enjoy. Feeding grain crops to animals and then eating their meat is less efficient than using the grain directly for human food.

Foods are primarily natural chemicals

All but about 140 lbs. of the 1,500 lbs. of food each of us eats each year comes directly from plants and animals. Sugar, salt, corn syrup, and dextrose account for 130 of these 140 lbs. Leavening agents and substances added to adjust the acidity or alkalinity of food products account for another 10 lbs. Some 1800 different food additives account for about one pound of our annual food intake. This includes contaminants which get into our food accidentally.

The chemistry of foods which come directly from animals and plants can be quite complex. For example, about 150 different molecules have been identified in potatoes. Forty-two kinds of molecules have been found in orange peels, and additional kinds of molecules are in the rest of the orange. Potatoes and oranges are no more complex than many other foods.

Not all of the substances in natural foods serve to maintain good health. The 50 kilograms (kg.) (110 lbs.) of potatoes an average person eats each year contain about 10 g. of solanine. This molecule is closely related to poisons produced in dangerously large quantities by the potato's relatives in the nightshade family. Ten grams of solanine is more than enough to kill a person if taken in a single dose.

The average person eats about one kilogram of lima beans each year. This quantity of lima beans contains 0.04 gram of hydrogen cyanide, the lethal agent used in gas chambers. A person's annual consumption of seafood includes 0.014 gram of arsenic. The two teaspoons of nutmeg an average person eats each year contain 0.044 gram of a strong hallucinogenic drug, myristicin.

But potentially toxic substances in foods are hazardous only when consumed in extraordinarily large quantities in a short time. Our body chemistry tends to get rid of small quantities of toxic substances in our daily food. Even if we ate one one-hundredth of the lethal dose of 100 different natural food toxins, the mixture would be harmless. In fact, in some instances, the toxicity of one molecule is reduced or inhibited by the presence of another toxic molecule.

People have learned to select and prepare food so that they are not exposed to toxic substances. There are exceptions, of course, amateurs who gather wild mushrooms, for example. But much has yet to be learned about the kinds of molecules which make up or are found in our food and how these molecules interact with each other and with other kinds of molecules which are added to or find their way into our food.

Everybody's food chemistry is unique

In many ways, everyone has the same body chemistry. Blood transfusions and organ transplants have proved dramatically that one person's molecules or cells and tissues, if very carefully matched, can be used without rejection by another person.

Transplanting hearts, kidneys, and other organs and transfusing blood without rejection becomes all the more amazing when we realize that the donor and the receiver may have quite different eating habits. Apparently body chemistry can take apart widely different kinds of "raw material" molecules and still put together precisely the kinds of molecules each person needs to maintain life.

Imagine, for example, a thousand people sitting down to a banquet. The menu includes many different kinds of molecules. If we assume that everyone eats the full menu, all of the takings apart going on in the digestive systems of all of the people could be the same. So could be all of the puttings together of new molecules.

But the chemistry of nutrition is not this simple. There could be wide differences in nutritional needs among the thousand people. Kinds of molecules that are in short supply in one person may be overly abundant in another person. Some people could have brought with them supplies of molecules which would react in an unanticipated way when they encounter certain molecules on the menu.

Also, each person has a unique set of genes, assuming that no identical twins are among our imaginary banquet guests. On this basis, the body chemistry of each person is controlled by a different set of "master molecules". Consequently, the environment and the materials with which the food being eaten must interact are different for each person. It would be foolish to claim that any one food would be equally good or bad for everybody.

It is truly amazing that all of the people at our imaginary banquet can eat so many different kinds of food and still maintain the harmony and rhythm of body chemistry that is essential to good health.

To keep food coming to our tables requires effort

There is a story behind each of the 8000 or so different food items awaiting us in a well-stocked supermarket. Each story begins where water, soil, and the sun's light and heat enable plants to grow. Plants build the substances absorbed from air and soil into the many kinds of molecules which form the tissues of roots, stems, leaves, flowers, seeds, and fruits.

Most important to us, plants produce food which is stored, as it were, between generations—food we can divert to fulfill our needs rather than those of the plants. Furthermore, plants and plant products can be fed to animals, and the animals, in turn, take apart the molecules that were put together by plants and use the pieces to build the molecules they need for their cells, tissues, and organs. Here again, we can divert the flesh and fat, eggs and milk of food-producing animals to meet our own needs.

Energizing the whole food-producing activity is the flow of the sun's energy through our environment. For reasons that have always stretched the minds and imaginations of people, the sun's energy is trapped in the complex molecules which green plants put together. This same energy is released when these molecules are taken apart in the bodies of animals. Not until the life processes within living systems convert this energy into heat is it permitted to continue its relentless flow toward entropy, the mysterious "final resting place" for all energy.

In somewhat less philosophical language, energy is the "glue" which can build complex molecules from atoms or from fragments of less complex molecules. The release of energy maintains life in the intricate systems of cells, tissues, and organs of plants and animals.

Three grain crops (wheat, rice, and corn) and three tubers (potatoes, yams, and cassova) presently provide four-fifths of the world's food energy.

To illustrate how plants produce our food and to show what it takes to keep food coming to our tables, suppose you decide to grow a year's supply of one of these six foods, potatoes. An average person in the U.S. eats about 240 potatoes each year.

Approximately 1,400,000 acres of farm land were used to grow the 13 million tons of potatoes that were produced in the U.S. last year. On this basis, you will need a garden about the size of an average living room. You will need to make sure that the soil in your garden contains adequate amounts of potassium, nitrogen, phosphorus, and at least traces of several additional elements. We assume that the rains will provide adequate water and that the sun will keep the soil temperature well above freezing and provide plenty of light throughout the growing season.

Now you need about 12 potatoes for "seed." Obviously, you will have to call on other "farmers" who raised more potatoes last year than they sold or ate. Similarly, these people had to rely on potatoes that were left over from the previous year's crop for the seed that yielded these potatoes. The point is, to be able to plant your 12 potatoes, someone somewhere along the stream of civilization had to discover the potato as one of nature's plants that could provide the food we need.

Historians tell us that potatoes grow wild in South America, particularly in the Andes. The Indians of the Andes discovered potatoes as a food some 4000 years ago. Some 100 years after Columbus made his first voyage to America, other European explorers took potatoes back to Europe with them. As a food for Europeans, at first the potato didn't catch on too fast.

By 1776, however, the potato had become established as an important food plant. Early in the 1800s, potatoes had become the main food in

Each person has a unique set of "master molecules" and, thus, a unique personality. The enzymes these molecules produce, however, enable people to take apart widely different kinds of food to obtain the specific building blocks each person needs. If some kinds of enzymes are missing in a person, certain foods would be very difficult to digest.

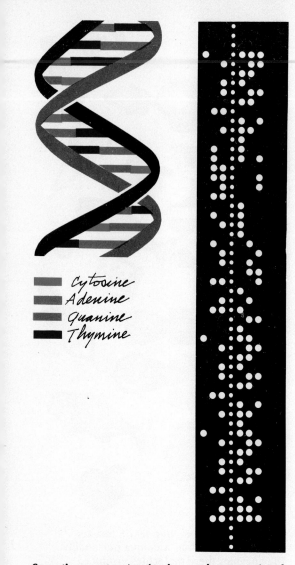

Cytosine
Adenine
Guanine
Thymine

Sometimes nature's miracles can be appreciated best by looking at them through the perspective of things we do understand. The computer tape is the "code" for the first sentence in Chapter 1. The DNA model is the "code" for a portion of the chemistry of a living system.

some countries, especially among the poor people of Ireland. These people ate as many as 20 to 30 potatoes each day. When a blight struck the potato crop in 1845 and 1846, wiping out nearly the entire crop, one and a half million people starved to death.

It is a simple matter to buy 12 seed potatoes at the nearest seed store. But tracing their ancestors would involve crossing the Atlantic ocean at least twice, and becoming involved in many historical activities.

Now that we have our seed potatoes, the next step is to cut each one into six or eight pieces. Each piece must have at least one "eye." The "eye" of a potato is the equivalent of a bud or sprout; the rest of the tuber is stored food.

Five or six equally spaced rows can now be marked out across our garden. Shallow ditches along each row will allow us to plant our 80 or so chunks of seed potato about 8 inches apart. After covering the seed with about 6 inches of loose soil, all we can do for the next few days is wait.

While we wait, in the smaller than microscopic world of molecules inside the cells of the potato seed, changes are taking place. Water in the soil slowly seeps through cell walls. Substances inside these cells slowly come apart. Heat from the sun finds its way through the soil and adds to the stresses occurring inside the cells. Large molecules called enzymes float among the other molecules. These enzymes bump into molecules of starch or protein and help break them apart. Soon the fluid inside the cells becomes a rich soup of bits and pieces of molecules, each carrying energy absorbed from the environment.

Other molecules on the scene are the genes that were packed into each cell when the seed potato was being put together by this same process. Now and then, a DNA molecule within a gene loses a fragment. This fragment floats around in the cell soup until it bumps into a similar fragment. Sooner or later, each fragment of the DNA molecule is duplicated. There are now two DNA molecules. When all the DNA molecules in all the genes that were present in an original cell have been duplicated, the cell will be under a new kind of stress—stress that can be relieved if the cell divides. Supposedly, each half can now absorb the materials needed to become a new cell from the surrounding fluids. This multiplication and enlargement of new cells is growth.

This process goes on within the cells of the seed potato until, after a few days, the tiny buds within the potato eye have grown enough to break through the soil and appear as sprouts above ground. Obviously, the starch and other complex molecules that were in the seed potato have been taken apart and their fragments built into the new molecules which make up the slowly developing roots and stem of a new potato plant.

Much of this story has been left out—Much of what happens when growth in the seed potato speeds up after the potato is planted is still pretty much mystery. Exactly how the stresses of added moisture and heat are relieved through cell division and growth can only boggle the minds of most people.

The author recalls being totally mystified by an unexpected new "crop" of potatoes that grew in a school science laboratory activity. A chunk of seed potato in a plastic bag with some water and vermiculite had remained overlooked in a lab bench drawer for several months. When finally discovered, the angular chunk of seed potato was only an empty potato peel. Still attached was a new small potato—small but shaped exactly like other potatoes. Its size suggested that nearly the total volume of the chunk of seed potato had been transferred to the new potato.

All but one of many attempts to reproduce this event have failed. Even so, it is a spectacular demonstration of how living systems respond to changes in their environments—changes which cause molecules to be taken apart and new molecules put together. These new molecules, together with the energy needed to form them, are our only hope for keeping the world's people fed.

Let's get back to our own potato "farm". Supposedly, by now leaves have appeared on our new potato plants. We can hope that no weeds have invaded our "farm" to rob the potato plants of the nutrients in the soil. We can also hope that no insects have found the new potato plants to be the food they need, and that no disease has invaded our rows of potatoes. Were we actual potato farmers with money invested in the crop and our income dependent upon harvesting a full yield of new potatoes, we would have to do more than merely hope that nothing would damage or destroy our new potato plants. This will be discussed later.

By now, all the food originally stored in the seed potato may have been changed to the cells and tissues of the new plant. But not necessarily. In the tissues of the new potato leaves are molecules of that well-known compound chlorophyll. Exactly how they got there is more of the mystery of things. But green chlorophyll is the trademark of the plant world. It serves a vital function.

Chlorophyll molecules trap the sun's energy as it passes through plant cells and tissues. As this energy builds up, new stresses are put on the fragments of molecules within the fluids in these cells and tissues, especially fragments of water and carbon dioxide molecules. As a result, the excess energy becomes tied into more complex molecules that are put together from water and carbon dioxide.

Soon the cell sap contains an increased concentration of these new energy-rich molecules, particularly molecules of glucose, $C_6H_{12}O_6$. In effect, six water molecules and six carbon dioxide molecules have been taken apart, and all but 12 of the oxygen atoms in these molecules have been used to put together a molecule of glucose.

Glucose production in the potato leaf cells continues as long as the sun is shining and carbon dioxide and water are being absorbed through the plant's roots and leaves. But as the newly formed glucose accumulates, new stresses appear in the cells. These stresses cause collisions between glucose molecules. Sometimes when the glucose molecules collide, a water molecule fragment splits off and two glucose molecules stick together. Apparently this relieves some of the stress caused by the accumulation of glucose molecules in the cell fluids, but only temporarily. Glucose molecules continue to collide, water molecules break loose, and more glucose molecules stick to each other.

Eventually large molecules of starch are formed. Because starch is not very soluble in water, the formation of starch molecules seems to relieve the stress of too many glucose molecules in the cell fluids. Starch molecules join to form starch grains. Soon the cell becomes packed almost full with insoluble starch. This marks the end of food-making in the cell. It is now a food storage cell, and cells forming elsewhere in the plant tissues continue the food-making process.

This is what happens in our potato plants while they are maturing during the summer. The genes of potato plants cause the food to be stored primarily in underground stems or tubers. The buds on the tubers become the potato eyes. The stems become "potato shape," because the cells are packed with starch. Other kinds of molecules, of course, form the skin and other tissues in the potato.

A typical potato contains mostly water. The starch and other carbohydrate contents runs from 17–34%. There may be 1–3% protein. As pointed out earlier, there are traces of many other kinds of molecules.

Again assuming good weather, not too many weeds, and no serious disease or insect damage, by the end of the summer we will find that each plant has produced several new potatoes from the chunk of seed potato we planted during early spring. We can also assume that we will dig these new potatoes before they are discovered by hungry rodents or other animal pests.

For our effort, we have the 240 potatoes we expect to eat during the coming year. There are still the problems of storage and processing, but before you lose your enthusiasm for being a potato farmer, let's assume these problems can be solved without too much additional effort and expense.

Our 240 potatoes solve a part of the problem of keeping us fed for a year. The chemistry of our bodies can put water molecules back into starch grains and individual starch molecules are set free. With the help of enzymes and additional water molecules, the starch is changed to glucose. Glucose is a soluble molecule which circulates easily through our body tissues and cells.

With oxygen available, the glucose in the cell fluids can be "burned" to release the energy we need. Our bodies have ways to get rid of the

In photosynthesis, that magician of the plant world, chlorophyll, and his energetic partner, the sun, team up to convert stable molecules from the soil and the air into an energy-rich building block and free oxygen.

Plucking carbon dioxide and water from the air and soil...

...and calling on his partner for a sprinkle of photons...

Zap.!! A carbohydrate building block and oxygen to replenish the atmosphere!

43

Food begins with the sun and soil…

water and carbon dioxide produced when the glucose is burned. This also gives us a glimpse of the elegance of nature's chemistry. It could be that the oxygen used to "burn" the glucose is the same oxygen given off when green plants put glucose molecules together out of fragments of carbon dioxide and water molecules. In turn, these raw material molecules could be the same ones produced when glucose is "burned."

The term "burned" calls for a bit of explanation. The glucose, for sure, reacts with oxygen. This is burning. We usually think of burning, however, as taking place at high temperatures—temperatures so high that cells and tissues would be damaged or destroyed. In living systems, complex molecules called enzymes enable burning, or the equivalent of burning, to occur at much lower temperatures. The same amount of energy is released no matter how slow or fast the process is.

In the language of school biology, photosynthetic processes in the presence of chlorophyll trap the sun's energy by building carbon dioxide and water molecules into complex molecules. In living systems, respiration reverses this process, making energy available and returning the water and carbon dioxide to the environment.

Fundamentally, food is chemistry

Food is chemistry, the chemistry of millions of different molecules interacting in rhythm and harmony. Each of these interactions must be repeated countless times. Each taking apart and putting together calls for specific raw materials, enzymes, catalysts, vitamins, or other maintenance

RESEARCH PROCESS MARKET BAKE RYE PACKAGE TRANSPORT

...but many people are needed along the path to the consumer.

molecules. Each is accompanied by waste materials which must be re-cycled or disposed of.

There is a unique chemistry at each stage of the growth of a food crop. One kind of chemistry predominates when seeds germinate or sprout. Other chemistry becomes more important when new seedlings are "weaned" from their food that was provided in seeds, and they establish photosynthetic independence. New sets of chemical processes predomi-nate when plants become mature, flower, set fruit, and when the fruit or seeds ripen.

Reduced to basics, the food production depends upon the providing of plants and animals with the essential building blocks and maintaining the environment that best supports the chemistry of life.

The unique role of research in farming

Modern farming benefits from two sources of knowledge. One of these is the experience that has been handed down through the thousands of generations of men and women who have matched their wits and hands against the challenge of growing crops and raising farm animals. The other is the knowledge that has been and is being revealed through the research efforts of agricultural scientists, engineers, and technicians.

"Dirt" farmers and "research agriculturists" probably are more alike than they are different. They both rely on painstaking observations. Farm-ers enjoy the advantage of extensive first-hand familiarity with their crops and farm animals. Day after day and year after year they observe the

45

Slight changes in the environment of a growing plant can produce visible effects. Here a healthy birch leaf (top) is compared with a leaf from a tree that had been exposed to air polluted with sulfur dioxide. The bottom photo shows a healthy Jerusalem cherry leaf compared with a leaf from the same tree exposed to air polluted with hydrogen fluoride.

subtle changes which show how normal and abnormal conditions affect the health and vigor of the plants and animals upon which they depend for their livelihood. Experienced herdsmen, for example, have an almost uncanny ability to sense that something is causing an animal to "go off its feed." And more often than not, these people know what to do to remedy the situation.

Successful gardeners and orchardmen can see in the subtle shades of green in vegetable or tree leaves an impending shortage of a specific plant nutrient or the initial symptoms of disease or insect infestation. They can see from the slight changes in color or firmness of fruit when they need to schedule all the chores that go with getting their products to market.

Agricultural scientists are equally good observers. In fact, little is to be gained by trying to compare "dirt" and "research" farmers. Both groups devote their knowledge and knowhow to the same basic goal: keeping food coming to our tables. If we seem to have more to say about "research" than "dirt" farmers, it is because the research group tends to publish more of its knowledge and knowhow.

Farming is very much a research-oriented enterprise. No farmer survives for very long unless he stays tuned sharply to the conditions which affect his crops and farm animals. Problems must be faced promptly, analyzed as thoroughly as possible, and probable hypotheses checked out until the problems are solved. Of course this is also true for research agriculturists.

Research aims at increasing the yield from farming

The basic business of farming is collecting and storing solar energy. This is done by encouraging green plants to combine carbon dioxide and water to make carbohydrates. We know that every ounce of carbon dioxide (about enough to fill two toy balloons) that can be made into carbohydrates will store 280 kilocalories (280 food calories). But farmers also know that some 30 times this much solar energy must fall on the corn, wheat, rice, or whatever crop is being used to store the 280 kilocalories. Increasing the efficiency of collecting and storing solar energy is a wide open challenge to the science and technology of farming.

More than 56 million acres of farm land in the United States are planted each year with soybeans. The annual yield is nearly 100 billion pounds, for an average per acre yield of about 1670 pounds. If some way could be found to increase this by only 1%, farmers would harvest an additional 935 million lbs. of soybeans. At an average price of $6 per bushel, this increased yield would be worth more than $93 million.

A 1% increase is not at all unrealistic. Keep in mind that corn yields were increased some 300% between 1950 and 1970. In fact, some farmers and research scientists are setting a goal of more than 4000 lbs. of soybeans per acre, and they expect to reach this goal. Here are a few glimpses of how some of these people expect to reach this goal.

James E. Harper knows that 400 lbs. of nitrogen would have to be absorbed per acre of soybeans to reach the 70-bushel-per-acre goal. But soybeans absorb nitrogen from the soil while the young plants are growing and then take nitrogen from the air as they approach maturity. In effect, fixation of atmospheric nitrogen by the bacteria which live on soybean roots competes with the absorption of nitrogen already in the soil. And nitrogen fixation can provide not much more than 100 lbs. of nitrogen per acre. Dr. Harper believes it should be possible to put together a new kind of nitrogen fertilizer molecule that wouldn't interfere with or inhibit the atmospheric nitrogen fixing process that goes on in legumes. Or maybe a strain of nitrogen-fixing bacteria can be found that would tolerate higher rates of soil nitrogen.

Richard L. Cooper knows that many varieties of soybeans topple over or lodge if their stems grow too tall. When soybeans lodge before they are ready for harvest, the highly organized arrangement of the green leaf area is damaged. Fewer leaves are able to collect their full quota of solar energy. Dr. Cooper is looking into the effects of planting soybeans in more closely spaced rows. Another hunch is that semidwarf varieties of soybeans can be developed, especially dwarf plants which would produce side buds and branches so they would not only keep each other from toppling over but also completely cover the soil with an energy-collecting surface.

For more than 10 years Fred W. Slife has headed a research team which believe that soybean yield can be increased by putting together better weed-killing molecules. These people hope to find herbicides which will eliminate weeds in soybean fields and possibly at the same time reduce nematode damage. Nematodes are tiny worms that feed on many kinds of crop plant roots and sap much of the plants' vigor.

William L. Ogren and Israel Zelitch are only two of the people attacking head-on the efficiency of the photochemical processes whereby green plants store solar energy. Dr. Zelitch is particularly aware that the first intermediate molecule put together by corn plants during photosynthesis contains four carbon atoms. The first intermediate molecule to appear in soybeans and other low efficiency plants has three carbon atoms per molecule. The question is why some plants start putting together the glucose molecule by way of 3-phosphoglyceric acid, the 3-carbon atom molecule, rather than the 4-carbon atom malate or aspartate.

Dr. Zelitch believes that research can find a kind of plant tissue with even higher solar energy-storing efficiency than that of any already known plant tissues. Now known are techniques which enable individual plant tissues to be cultured in bottles. By using this tissue culture technique, Dr. Zelitch believes that new information will be revealed to enable research people to spell out in greater detail, and hence manage more efficiently, the basic processes of photosynthesis.

Dr. Ogren's particular interest is in the carbohydrates put together in the soybean plant but taken apart later to provide the energy that is used to keep the soybean plant alive and growing. He believes an enzyme can be formulated to allow soybeans to collect and store a greater net fraction of the energy-rich carbohydrates that are put together during photosynthesis. He has treated soybean seeds with nuclear radiation and other agents known to change genes and chromosomes, in the hope of producing a new variety of soybeans in which the desired enzyme would appear. Adding zest to this challenge is the belief that soybean yields could be increased 50% if the plants consumed only as much of their newly made carbohydrates as they actually need.

Very subtle factors, however, control the balance between photosynthesis and respiration in the soybean plant. For sure, oxygen is being used to release energy all the time the plant is alive. This energy comes from the oxidation of complex molecules such as carbohydrates. But oxygen is made available the entire time carbohydrates are being put together from carbon dioxide and water. In other words, everything is present in the cells and tissues of green plants to allow carbohydrate molecules to be destroyed and constructed simultaneously. Supposedly, when exactly the proper amount of light is falling on the green plant and the proper concentrations of water and carbon dioxide are at hand, the plant continues simultaneously to live, grow, and store energy at a maximum rate.

For their research, Orlang A. Krober and Robert W. Rinne are investigating a step beyond photosynthesis. They believe it is possible to improve the efficiency of the soybean plant's changing of carbohydrates into fats and proteins, the kinds of molecules which actually end up storing much of the solar energy collected by the plants. They contend that if the biochemical pathways whereby fat and protein molecules are put together can be spelled out in sufficient detail, methods can be found to increase their efficiency.

Ten to 12% losses to disease create a challenge to the men and women who seek to increase farming efficiency, especially in raising soybeans. Most research in this area tries to find or breed disease-resistant varieties rather than to create chemicals which prevent or kill disease-causing organisms. Richard L. Bernard, for example, collected more than 2000 soybean samples in Japan and Korea in the hope that somewhere in all of this "germplasm" he might someday isolate the gene or genes that would bring resistance to one or more of the diseases which damage soybeans.

In related work, Dale I. Edwards believes that another cause of soybean damage, the soybean cyst nematode, can be controlled by blending seeds of resistant and susceptible varieties. His logic is that the parasites would be trapped in the susceptible variety and leave the main crop alone.

Finally, W. Ralph Nave illustrates how another group of people is seeking to improve agricultural efficiency by developing better harvesting procedures. About 12% of the soybeans produced in the U.S. are lost during harvest. High-speed photography is being used to see exactly how poorly-designed harvesting equipment fails to collect individual soybeans during harvest. Agricultural engineer Nave believes that the equipment can be redesigned to put a higher percentage of the soybeans being harvested in the bin.

These brief glimpses of some of the people who are trying to increase the yield of soybeans merely suggest how large a role research plays in farming. Similar glimpses of research on other kinds of farm crops and animals can be obtained in dozens of university, industrial, and government laboratories and experiment stations. Furthermore, the need to solve problems in order to keep food coming to our tables will continue to attract young men and women who are willing to match their wits against all the things and conditions which influence the health, vigor, and productivity of farm crops and animals.

The soil is a huge storehouse of building blocks

Farming depends on the soil. This is a blunt statement, almost a truism. But it is not easy to dramatize sufficiently how dependent on the soil we are for our food. The trouble is that supermarket shelves and freezer cabinets seem to be a long, long way from the red clay of Georgia, the black loam of Iowa, or the just plain dirt of the world's farms, orchards, and gardens.

To some people, the thin covering of soil over the earth may be nothing more than where the land leaves off and the air begins. Put in better perspective, the soil is the storehouse of the building blocks which are essential to putting together the food we need to survive. A quantity of soil no larger than can be piled on a teaspoon may contain more atoms of all of the earth's elements than there are people in the world—in fact, 250 billion times as many.

In *Biology and the Future of Man*, a book edited by Philip Handler and published in 1970 by the Oxford University Press, we can read, "A cubic inch of fertile soil can contain as many individual living organisms as there are people on the earth" (p. 588).

Furthermore, that spoonful of soil is the scene of infinite takings apart and puttings together. The soil organisms and chemical processes are constantly taking apart the complex molecules which are the debris of the world of life. The soil is a fantastically effective sewage and waste disposal system. Given time, virtually all organic molecules are taken apart and their building blocks made available to whatever puttings together are at hand. The materials of the earth are cycled and recycled as life goes through generation after generation.

Essential plant minerals must be returned to the soil

Adding nutrients to the soil to insure good yields of grains, fruits, and vegetables has become a necessary part of farming in nearly all of the world. In frontier days, the soil in much of the United States contained a store of plant nutrients that had been accumulating for centuries. Even so, early colonists soon learned the advantages to be gained from returning manure to the soil. And there is the often-told story of the Indians teaching the colonists to bury fish in the hills where corn was planted.

For reasons which are difficult to trace, some people worry about using fertilizer which does not come directly from organic sources, that is, from wastes produced by plants or animals. It is interesting to see how this worry was handled by a student who chose to look into the chemistry of farm fertilization in response to an assignment involving the role of chemistry in modern living. In fact, his total discussion of the topic of fertilizers tells us much of what people, particularly young people, can learn about this part of farming. His paper is presented here with only minor editing.

Farmers usually try to maintain high levels of plant nutrients by natural means: by rotation to different fields of certain crops that require specific nutrients, by planting crops such as clover or alfalfa that help to replenish the soil every three or four years, and by plowing under hay, straw, or cornstalks in order to help speed up bacterial decay and increase the amount of soluble nutrients.

Often it becomes necessary to add these soluble nutrients to the soil so they can be used directly by plants. The three elements most often added to the soil are nitrogen, phosphorus, and potassium. Nitrogen promotes vigorous growth and better foliage in most plants except the legumes, which produce approximately two-thirds of their own nitrogen. There are no nitrogen-containing minerals in the soil, and plants either fix nitrogen from the air or receive their nitrogen from ammonia, nitrates, or nitrites.

Nitrogen is added to the soil by commercial fertilizers made from dried blood, meat and fish scraps, sodium or calcium nitrate, ammonium nitrate or sulfate, liquid ammonia, or other similar compounds. The manure from livestock also provides nitrogen for crops.

Each year millions of tons of nitrogen are produced by natural nitrogen fixation during thunderstorms. Lightning causes the nitrogen in the air to combine with oxygen to form nitrogen oxides. These oxides, in turn, dissolve in water to form nitric or nitrous acid. These acids react with minerals in the soil to form soluble compounds which make nitrogen available to plants.

Another way to obtain usable nitrogen is by making calcium cyanamid. Nitrogen from liquid air is passed over white-hot calcium carbide. The cyanamid can be used directly or converted to ammonia by means of superheated steam. Many farmers now spray 40–80% ammonia onto superphosphate contained in a revolving mixer and then apply this mixture to their crops.

Potassium is especially needed by plants that produce large amounts of carbohydrates. Most raw materials for potassium fertilizer are mined in the form of crude carnallite which is made into more soluble potassium chloride by treating it with a hot solution of 20% magnesium chloride. Potassium salts are also obtained from lake brines which are also used to produce ordinary table salt by evaporation.

In some ways, phosphorus compounds are the most important plant fertilizers. Phosphorus is needed, for example, in the ripening processes of grains. Phosphate rocks in nature are very insoluble, particularly calcium phosphate. But if

Nutrients in the soil enable plants to grow and to make the food they store. Depleted soil produces stunted plants. Nature's processes recycle and restore the soil's supply of nutrients. Farmers lend nature a hand by plowing under soil-building crops and by fertilizing their fields. Formerly depleted soil can thus be made to yield good crops.

the phosphate rock is treated with sulfuric acid, hydrogen atoms join the calcium atoms in the calcium phosphate and a soluble product is formed. A more concentrated superphosphate is formed if phosphoric rather than sulfuric acid is used. Nitric acid can be used also; then the finished product will provide both phosphorus and nitrogen.

All mixed fertilizers are labeled to show their content. A 4-12-5 fertilizer, for example, would contain 4% nitrogen, 12% phosphorus, and 5% potassium.

Campaigns against commercial fertilizers have been started by some people. Environmentalists sometimes claim that these fertilizers ruin drinking water and kill many forms of aquatic life when the runoff from farms reaches lakes and streams. To ban these fertilizers before better forms of plant nutrients are available, however, would result in ruining the soil, as with the 1933 "dust bowl." Crop yields would drop disastrously.

Having worked the last few summers farming with my grandfather near Dekalb, Illinois, I get the general feeling that farmers actually know very little about the chemistry of the fertilizers they use. Fertilizing traditions seem to have been passed down through the generations and many appear to be antiquated. I believe if farmers learn to analyze their own soil, they can avoid wasting a great deal of fertilizer and still not deplete the soil. Having applied ammonia with a type of plow myself, I believe much fertilizer could be saved by better methods of application. And chemistry seems to hold most of the solutions to current fertilizer problems.

Small quantities of some kinds of building blocks in the soil may play large roles in maintaining the health of plants, animals, and people

The roles of nitrogen, phosphorus, and potassium in keeping plants growing vigorously are well known, as is the role of calcium in not only providing this building block but also in helping to keep the acidity or alkalinity of soils in balance. Several other elements are important in the production of our food, but their roles are less clearly understood. For one reason, these elements exist in concentrations so low that they are reported to be present only in trace quantities.

Some of these trace elements are believed to help maintain good health in plants and animals; others are harmful. Fortunately, plants provide effective barriers against several trace elements present in some soils that could poison animals if they were taken up and deposited in plant tissues. These include arsenic, iodine, beryllium, fluorine, nickel, and vanadium. On the other hand, plants can grow normally and still pick up enough selenium, cadmium, molybdenum, or lead to be harmful in animal feeds or human foods.

Also, plants can grow in soils having low concentrations of some of the trace elements which are essential to good health. Foods from these plants may fail to meet dietary requirements because they contain too little cobalt, chromium, copper, iodine, manganese, selenium, or zinc. Deficiencies of trace elements create problems which have been known for many years. There is a story, for example, about an area in New Hampshire where crops can grow; but when these crops are used for cattle feed, the cattle become weak, anemic, and emaciated. Many die. A local legend says the soil carries the curse of an Indian chief who resented seeing his land taken over by the colonists. Today we know that the problem is really a shortage of cobalt in the soil. To solve the problem, animals are given blocks of salt containing cobalt to lick and trace quantities of cobalt are added to the fertilizers used in the area.

Some soils contain plenty of iron, but the iron atoms are tied up in molecules in such a way that they are not absorbed by plants. In the problem of "tired blood," some people can't obtain the iron they need even though food crops contain iron. Darrell Van Campen has studied this problem. He believes that some amino acids, especially histidine and lysine, help make iron available when it is present in foods.

When iron alone was injected into rats, only 5% of the iron was absorbed in three hours. When lysine was added to the iron solution, about 50% of the iron was absorbed. Histidine was even more effective. Further research may suggest that small quantities of these amino acids added to foods, along with soluble iron compounds, might prevent or cure iron-deficiency anemia.

$H_2N(CH_2)_4CH(NH_2)CO_2H$

lysine

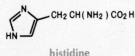

histidine

50

Zinc, like iron, can be present in soil but not in compounds which make this building block available in our food. Zinc deficiency in plants is becoming more widespread, and consequently less available in food, especially plant proteins. Zinc plays a role in healing of wounds and surgery, in animal reproduction, and in counteracting the effects of cadmium poisoning. Jean Apgar found that rats raised on soybeans containing zinc in unavailable form had difficulty giving birth. Furthermore offspring born to rats fed zinc-deficient diets during pregnancy died shortly after birth. If available zinc were added to their diets, the rats delivered their offspring normally and raised them successfully.

Harold H. Sandstead and Edward S. Halas have also studied the effects of too little zinc in the diets of animals. They found that feeding zinc-poor diets to pregnant rats caused their offspring to learn less rapidly than those of rats fed adequate zinc. These offspring failed to grow at normal rates, were slow to mature sexually, seemed to lose their senses of taste and smell, and healed slowly.

Farm crops can be grown in artificial or synthetic soils

That plants will grow in the absence of soil has been known ever since John Woodward in 1697 found that plants would grow in polluted river water without soil. In the 1850s, Jean Boussinggault grew plants in sand to which water containing chemical nutrients was added. In 1860, Julius von Sachs published a formula for the chemicals needed to grow plants hydroponically, that is, without soil.

Modern hydroponic projects produce yields of special interest to the people who worry about whether the earth can continue to feed the world's population regardless of how rapidly people reproduce. Hydroponic set-ups in India, for example, double the yield of tomatoes, lettuce, and beets, and increase the yield of rice from 900–5000 lbs. per acre.

In water-culture hydroponic "gardens," the plants are grown in nutrient solutions without even sand or gravel as a supporting medium. Before planting, the seeds are germinated and then placed on cork floats so they are held slightly above the nutrient solution. In large scale productions, the seedlings are planted in peat moss or wood shavings held above the nutrient solution by a net of asphalt coated chickenwire.

Because of the high cost of growing crops hydroponically, this method is limited to areas where environmental conditions do not permit crops to be grown outdoors in natural soil. But some day hydroponic farming may be the only way to feed the world's people.

Chemistry, biology, and physics all lend a hand to protecting farm crops and animals from insects

Putting together molecules that help farmers control weeds, insects, rodents, and other pests is a longstanding and immensely demanding challenge to chemical science and technology. Gallons of printer's ink have already been spread to tell the story of the apparent conflict of interests between the people who are responsible for keeping food coming to our tables and equally dedicated people who are worried about our environment. These people see the widespread use of pesticides as threats to the health and well-being of all plants and animals.

This conflict of interests is very, very complex. Farmers are well aware of the delicate balance existing among the living systems in an environment. They are equally aware of the dependence of living systems on rather narrow changes in temperature, moisture, and other environmental factors. They are all too familiar with widespread "crop failures" and livestock epidemics when unfavorable conditions or sudden "population explosions" of disease-producing organisms sweep through fields or feedlots.

Farmers know very well that a favorable environment cannot be taken for granted. An environment in which their crops and farm animals can flourish must be achieved through long-range planning and careful management.

Weeds and insects have threatened our food supply for a long, long

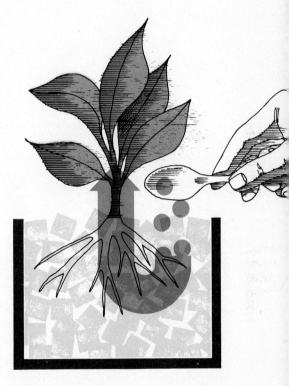

Successful hydroponic farmers provide precisely measured amounts of all of the chemical nutrients plants need.

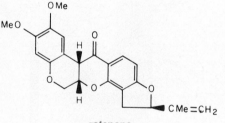

rotenone

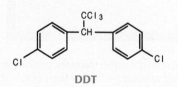

DDT

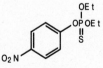

parathion

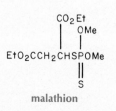

malathion

H_2NCO_2-

carbamate

time. Ravages of insects during Biblical times are described in the book of Joel as, "What the locust has left the swarm eats, what the swarm has left the hopper eats, and what the hopper has left the grub eats."

In the United States, crop damage by insects adds about $20 a year to each person's food bill. In some parts of the world, probably half of the food produced is consumed by insects, rodents, birds, and other "enemies of the people."

In the early days of farming, pests usually were removed from crops by hand. Many people have childhood memories of chopping weeds out of cotton fields; hoeing long, long rows of corn; or carrying buckets of lamp oil along potato rows and picking off the fat, chocolate-red larvae of the Colorado potato beetle and dropping them into the lamp oil. Each adult beetle with its bright yellow and black stripes was a special find because we knew well that this was the source of the chewed up leaves and stunted plants.

Then came lead arsenate, Paris green, sulfur compounds, nicotine, and other highly poisonous materials for sprays and dusts. These, too, have memories—memories of strict precautions against leaving these poisons where unsuspecting people, especially children, might use them inadvertently. There was always the worry lest favorite farm animals or pets would eat either the crops which had been freshly sprayed or rodents that carried lethal doses of poison.

Childhood memories also include sheets and strips of sticky fly paper left on flat surfaces or dangling from ceilings that always seemed to be much more intent on catching elbows and spoiling hair-styles than trapping houseflies, and shallow dishes containing pieces of black "poison fly paper" soaking in water—water often as attractive to toddlers as to houseflies.

Then came new brands of insecticides with labels saying they were quite poisonous to insects but much less so to people and farm animals. Rotenone and pyrethrum, although more expensive than inorganic poisons, worked well. It took time, however, for people to really believe that substances could kill pests and not people or other animals. How substances which could kill pests without killing people puzzled farmers until they realized that there are plants which insects will not touch; hence, there are plants which produce their own insecticides.

If plants can put together molecules which kill bugs, then people should be able to put together molecules which kill bugs and not people. In retrospect, this kind of thinking may have been going on in the minds of certain people during the 1930s or even earlier. In 1874, for example, Zeidler in Germany put together a molecule in which chlorine atoms were bonded to molecules containing carbon atoms. One of these "chlorinated organics" is dichlorodiphenyltrichloroethane, much more commonly called DDT.

In the 1920s when people in Germany were looking for better moth-proofing agents, they realized that the DDT type of compounds might be what they were looking for. The insect killing effectiveness of DDT was discovered by Paul Muller in Switzerland in 1939, a discovery for which he received the Nobel Award in 1948.

More about the story of DDT appears in Chapter 7. It is particularly thought-provoking to realize how much controversy there was during the 1950s and '60s involving a molecule that brought Muller a Nobel Award in the 1940s.

In the 1940s DDT brought to mind DC-3's flying over mosquito-infested countryside leaving behind trailing fogs of DDT mist, and GI's pumping dusters sometimes too enthusiastically under the clothing of recently liberated people in countries where typhus-transmitting lice were as much a threat as were enemy bullets. Thousands of World War II "tourists" could now "drink the water and eat the food" of war-devastated countries because DDT killed the houseflies that spread gastrointestinal diseases. Military personnel who were stationed where malaria was a constant threat knew that this threat became less serious after the countryside had been sprayed with DDT.

However, these were not the images associated with DDT in the 1950s and 1960s. By then other stories were being dramatized in newspapers and newscasts and in TV documentaries. These were stories of DDT residues decreasing the population of desirable species of animals and up-

In the past, either insects took a large bite out of our food or fighting insects took a large bite out of our free time. Even today insects consume approximately 12% of the food farmers produce.

setting the delicate balance of nature in all kinds of environments. The threat of increasing concentrations of DDT residues in the cells and tissues of people generated widespread uneasiness—uneasiness because no one knew for sure what longterm effects DDT might have on our health and well-being.

One effect of this uneasiness was to cause chemists to try to put together new molecules that would kill insects but not have the possible side effects which worried so many people. In 1938, Gerhard Schrader put together the molecule tetraethyl pyrophosphate. This molecule proved to be as toxic to insects as nicotine and other established insecticides. The discovery led to a group of new insecticides, the organophosphorus compounds. Soon several of these insecticides, parathion and malathion, for example, filled in for DDT.

These new insecticides did not linger in the soil and on vegetation nearly as long as did DDT, and thereby avoided the hazard of excessive residue buildup. They were used widely not only to increase crop yield but also to control animal pests such as cattle grubs, horn flies, fleas, and ticks. These insecticides entered animals' circulatory systems and provided long-term protection against pests. One popular method of application was to imbed the insecticide in plastic such as the flea collars that were sold for cats and dogs.

In the late 1940s, Swiss chemists put together a series of molecules suggested by a substance that had been found in the poisonous Calabar bean. These molecules became a series of insecticides which proved effective against many kinds of insects. Slight variations in the makeup of the carbamate molecule yielded insecticides highly effective against specific insects. This became a special advantage in that a pest could be controlled with minimum threat to other species in the environment.

Food is many things to many people

Eating is a personal matter. What we eat and how much we eat is pretty much up to the individual. This is really the point at which nutrition problems begin; and at which they must be solved.

How much of our day we spend feeding ourselves differs widely from country to country and depends on whether we are poor or affluent, whether we prepare our food from scratch or rely on thaw and serve. Some people keep a full pantry stocked with much of the food they will need weeks or even months ahead. For other people, it is "hand to mouth."

Our need for food is so basic that food is an important factor in family structure, in sociability and entertaining, in religion, and in the fulfillment of the urge to be creative. People enjoy growing things, especially prize animals or bumper crops of attractive agricultural products. Tradespeople take pride in offering us artistic arrays of fruits and vegetables, and it is a rare cook or chef who does not take pride in preparing and serving food.

Food is both an enterprise and an industry
We need at least a minimum amount of food

Today, the amount of food that is available for the world's people can be a very soul-searching topic. For 700 million people, overeating is as much a problem as is hunger. Another 700 million people obtain just about the amount of food they need. Millions of these latter people, however, have food problems because they sometimes run out of money or simply don't know the kinds and amounts of food a person needs.

For more than two billion people, where the next meal is coming from is on their minds constantly. All too many of these people survive on a margin of life so degrading as to insult human dignity. Only rarely do these people have the meals which provide all of the building blocks needed to maintain the rhythm and harmony of a well fed body.

During the first millions of years there were people on earth, it is believed they lived in small groups and obtained their food by hunting and gathering plants and animals in the area. About 10,000 years ago people began cultivating plants and domesticating wild animals for food. Exactly why and when people became farmers is a story that can be put together only by using archeological evidence. Different archeological digs yield different kinds of evidence; hence, there are different versions of the story.

But all efforts to describe the origin of agriculture have one thing in common. People have always been aware of their dependence upon nature for their existence and they have known for a long time that they could lend nature a hand in producing the food they needed. It is this awareness that has always fed mankind. Solutions to the problems of world hunger are most likely to be found in our becoming increasingly aware of our dependence on nature's chemistry for existence, and in backing up this awareness by lending nature a more effective hand in producing and distributing the food we need.

4 Around Us: What Chemists Make for People

Examples of materials created to have the properties needed to solve specific problems range all the way from wet suits for deep sea divers to special plastics and fibers which protect astronauts from very hostile environments.

Chemists make the materials we need

Our material needs are met by matching the characteristics or properties of substances with our demands. The properties of materials, in turn, depend upon what they are made of and how their building blocks are put together. If you change or rearrange the atoms in the molecules of a substance, for example, the properties of the substance will change.

Some of the things chemists make are exciting while others are commonplace. Some of the stories behind chemical progress involve only minor changes in the raw material's molecules; others require the joining of bits and pieces of molecules from many different kinds of raw materials. In some cases, the rearrangement of atoms occurs almost automatically when the raw materials are brought together; in other cases the reaction must be carefully controlled.

Glass that is hard to break

Glass is useful, but it breaks. The story of unbreakable glass illustrates how chemists can improve the properties of materials.

In 1961, S. S. Kistler, a professor at the University of Utah, announced that he had found a way to strengthen glass. Dr. Kistler's idea involved chemically tempering glass. Tempered glass is somewhat like a curved board's being used as a footbridge across a stream. If the board is put under stress by curving it upward and if people walk on the upper curved surface, the board will be less likely to break under their weight. If the outer layers of glass can be similarly stressed, the glass is less likely to shatter when it is hit or subjected to shock.

Dr. Kistler's idea was to stuff the surface of glass by exchanging larger atoms for smaller atoms in the glass surfaces. This way the glass structure near the surface is more crowded and thereby produces the compressive stress needed to strengthen the glass.

Soda-lime glass is made by melting a mixture of soda ash, lime, and sand. Other names for these raw materials are sodium carbonate, Na_2CO_3, calcium oxide, CaO, and silicon dioxide, SiO_2. If a batch of these raw materials is heated hot enough, clear, transparent glass forms. The structure of glass is a complex arrangement of sodium, calcium, and silicate ions.

Note particularly the sodium ions that are included in the glass. Sodium and potassium ions share many properties. Both elements are alkali metals, but potassium ions are larger than sodium ions. Therefore, if potassium ions can replace part of the sodium ions in the glass surface particles, the structure will be forced into a condition of compression.

A group of scientists at Corning Glass worked out a way to "stuff" potassium ions into the surface of glass, especially the glass that is used for eyeglass lenses, aircraft windshields, and laboratory ware. Actually, it is easier than you may think to exchange the larger potassium ions for the smaller sodium ions. According to the kinetic-molecular theory, all of the ions in the glass particles are vibrating back and forth constantly, and the degree of motion can be increased by raising the temperature of the

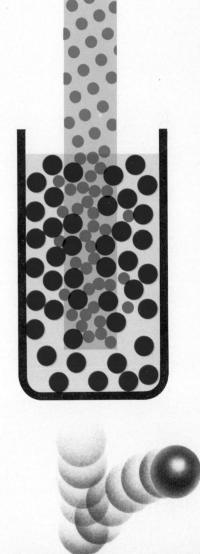

To chemists, whether or not glass breaks depends upon how its building blocks are put together. Chemtempered glass doesn't break because sodium ions have been replaced by potassium ions.

glass. True, none of this motion can be seen, but it can be taken advantage of to bring about the required transfer of potassium for sodium ions.

To bring about the exchange, potassium nitrate, KNO_3, is heated to its melting point in a stainless steel tank. At this high temperature relatively large gaps occur between the potassium and nitrate ions during each vibration. The glass that is to be tempered is then lowered into the molten potassium nitrate. As the temperature of the glass rises, the vibration of the ions in its particles also increases.

Under these conditions, it is easy to see how potassium ions become trapped here and there in spaces recently vacated by sodium ions. After being immersed in the molten potassium nitrate for approximately 16 hours, the outer layers of the glass become stuffed with enough potassium ions to create the compressive stress required for tempering. Stuffing of the outer layers of glass also causes the core glass to be in tension.

The glass is now taken from the molten potassium nitrate, cooled, washed, dried, and is ready to use. Objects which hit this glass will bounce off just as the bent board supports the weight of people crossing the footbridge. Tempering has improved the strength of this material. The glass is less likely to break because the compressive layer must first be penetrated. It is still glass, but it is better glass.

This better glass holds promise of solving additional problems. For example, the people who work at the research and development organization called Survival Technology believe that many of the victims of heart attacks could be saved if their conditions could be diagnosed promptly and the proper medicines given immediately. They have developed a pocket-sized device which enables a person who has a history of heart attacks to transmit an electrocardiogram by telephone to a hospital's emergency room. The patient is also to be equipped with syringes so he can immediately inject the needed medicines as revealed by the "over-the-phone" cardiogram.

One problem in this system is providing patients with prefilled syringes which are reasonably unbreakable and which are satisfactory containers for heart condition medicines. This is where tempered glass comes in. Tempered glass has the properties needed to solve these problems. This also shows how new inventions depend quite often on the availability of materials with unique properties.

Glass that pulls down the shades

Glass is often used simply because it passes light. Sometimes, however, new problems arise if too much light passes through. The discomfort caused by facing bright sunlight is a good example. In extreme cases, we can be blinded by too much light hitting our eyes. The hazards of glaring sunlight are very real to automobile drivers and airplane pilots.

Dr. S. D. Stookey and Dr. W. H. Armistead describe a way to improve glass so that it automatically adjusts the amount of light that can pass through. Their method takes advantage of several facts and concepts. For one, they knew that molten glass dissolves many substances—in fact, molten glass comes closer to being a universal solvent than does water.

When something dissolves in molten glass, the dissolved material is distributed throughout the glass structure. If silver chloride, AgCl, is dissolved in glass, for example, the silver and chloride ions can be pictured "as fish swimming through a three-dimensional net whose meshes become gradually more rigid and smaller in size as the material cools," using Dr. Stookey's words.

People who have worked with photography are familiar with another property of silver chloride. This is one of the materials that can be used to make photographic film. Several silver compounds darken when exposed to light.

Putting such facts as these together yields a possible solution to the problem of too much light coming through glass. Silver chloride can be dissolved in molten glass. The molten glass is cooled carefully, and the silver chloride forms very small crystals distributed uniformly throughout the glass. If the silver and chloride ions are in very small crystals, they allow light to pass through the glass.

Silver and chloride ions in crystals are bonded to each other by sharing a pair of electrons, and electrons are very similar to photons, the tiniest possible bundles of light energy. If we think of light as being a stream of photons, it is easy to imagine that light affects the bonds between silver and chloride ions. Thus, when light passes through the glass, the silver ions regain the electrons they have been sharing with chloride ions and become silver atoms. Silver atoms interfere with light passing through the glass.

This is similar to the chemistry of photographic film. Unexposed film is covered with layers of silver salts. When the camera shutter clicks open and light is focused on the layers of silver ions, the silver ions are started on their way to becoming silver atoms. The dark portions of a photographic negative are the result of varying concentrations of silver atoms that have been "developed" and left behind after the unexposed portions of the film have been dissolved and washed away.

However, silver ions dissolved in glass, sometimes called photochromic glass, behave somewhat differently. Unless the stream of photons continues to pass through the glass, the silver atoms return to share their electrons with the chloride ions, and the glass becomes transparent again. This is why photochromic glass solves the kinds of problems it does. Sunglasses are a good example. If the lenses are made from photochromic glass, bright sunlight changes the dissolved silver ions to silver atoms. The glass darkens and eyes are protected from too bright light. When too much light is no longer a problem, the silver atoms change back to silver ions and the glass regains its transparency.

Usually chemists are most concerned with the properties of whole atoms or molecules. Photochromic glass, however, illustrates how a problem can be solved by controlling the exchange of pieces of atoms—in this case electrons—between atoms.

Lithium, a behind-the-scenes building block

The usefulness of lithium depends more upon how it affects the properties of compounds than upon its properties as a pure element. Hence the expression: a behind-the-scenes element. Only rarely does lithium appear "front and center," but like stage hands, prompters, and others who work backstage, lithium has important supporting roles.

Lithium plays a role in making heat-resistant, glass-ceramic cooking ware. It is partly responsible for the color and ease of care of porcelain enamels. It is used to make laundry bleach and compounds used in swimming pools to keep the water in good condition.

In the medical field, lithium is used in making vitamin A, antihistamines, tranquilizers, oral contraceptives, and anticoagulants for blood samples. Industrially, lithium compounds are used in air conditioning and dehumidifying systems. Nearly half of the lubricating greases used in the United States contain lithium compounds. These greases hold up over wide temperature changes.

Lithium is used to control the production of some kinds of synthetic rubber, and it played a role in putting together the first rubber that duplicated exactly the properties of natural rubber. When added to aluminum, lithium produces an alloy with the properties needed to solve many of the problems encountered in metallurgy. The wing skins of high speed aircraft, for example, owe their excellent resistance to heat and corrosion to the properties lithium contributes to the lithium-aluminum alloy.

Lithium compounds are used to absorb the carbon dioxide in the air in the crew's quarters in submarines. A similar role is played in controlling the environment in space vehicles. The ready market for lithium supports the industry that finds sources of the element, extracts it from its ores, and makes it available commercially.

One source of lithium is the ore, spodumene, $LiAlSi_2O_6$, a lithium-aluminum silicate that contains about 8 percent lithium oxide, Li_2O. One ton of this ore yields about 70 kilograms of lithium. Deposits of spodumene are found in North Carolina. The ore is taken from the mine and crushed. Other kinds of rocks that are mixed with the spodumene are removed. The spodumene concentrate is heated and ground to a fine powder. Hot sulfuric acid, H_2SO_4, converts the lithium to lithium sulfate, Li_2SO_4, which is dissolved in water.

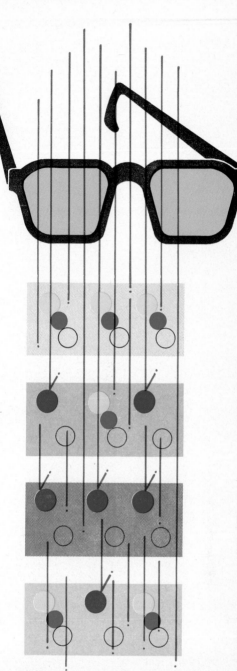

The more light that falls on photochromic glass, the less transparent the glass becomes. Light changes transparent silver ions to opaque silver atoms. When the light dims, the silver atoms change back to silver ions.

All of the insoluble portions of the original ore are now filtered off. The lithium sulfate solution, however, contains impurities which must be removed. To do this, sodium carbonate, Na_2CO_3, is added to the solution until lithium carbonate, Li_2CO_3, crystallizes in reasonably pure form. The crystals are harvested and then treated with hydrochloric acid, HCl. Carbon dioxide bubbles away, and a solution of lithium chloride, LiCl, is formed. By evaporating the water, crystalline lithium chloride is collected and dried.

To obtain pure lithium, the dry lithium chloride is mixed with potassium chloride, KCl, and heated until the mixture melts. The high temperature causes the lithium chloride to separate into lithium and chloride ions. When electricity is brought into the molten mixture, the lithium ions are attracted to the electrode where there is a supply of electrons. The lithium ions become atoms, and lithium metal collects on the electrode. The electrode is shaped so that the metal can be dipped out, allowed to solidify, and prepared for market.

It is interesting to question why lithium atoms contribute to compounds the properties they do. Such questions take us into the invisible realm of atomic structure where we must deal with mental models— mental models which we put together from facts we know and ideas we create. We know that lithium atoms are put together from three protons, three electrons, and four neutrons. Two of the electrons are pretty much "locked up" in a sphere around the proton-neutron nucleus. The third electron is left by itself and somewhat shielded from the positive electrical forces which hold it to the nucleus, or at least keep it in the neighborhood of the nucleus.

It isn't easy to see how a loosely held electron has much to do with the properties of ceramics, vitamins, medicines, alloys, and all of the other materials in which lithium plays a role. However, one of the basic assumptions of chemistry is that atoms stick to each other because their positively charged nuclei attract each other's outermost electrons.

The Great Salt Lake in Utah contains an estimated 4 million tons of lithium chloride. Because of the quantities of this building block that are available and because it has proved to be amazingly useful, lithium continues to attract the interest of many research chemists.

Chemistry produces many cosmetics and toiletries

The urge to beautify our skins is almost as old as civilization itself. Even the most primitive people have used the juices of brightly colored berries or pastes made from colored clays to decorate their skins. Ointment and cosmetic jars are found often in archeological digs. The earliest literature tells of women darkening their eyelashes and brows with a paste made from antimony, soot, and galena, PbS, an ore of lead. Green eye shadow was made from malachite, $Cu_2(CO_3)(OH)_2$, a copper ore. Cleopatra's use of eye shadow, paints, and powders and her tinted nails and palms probably measured up rather well to Hollywood's version.

In the early days of America, except for small quantities of imported perfumes, cosmetics were made at home. Pomades and ointments were made from lard, bear's fat, goose grease, or beef and mutton tallow. Home grown herbs such as lavender or thyme provided perfume. The household flour barrel solved the problem of too many freckles. Garden beets reinforced the color of lips and cheeks. For rough hands, there was mutton tallow, an excellent source of lanolin which is known today to be a useful ingredient in hair and skin care products. Burned corks solved the problem of darkening eyelashes.

Cosmetics today are the products of a complex blending of the knowledge and skills of chemists, biologists, physicists, and health scientists. Hit-and-miss or rule-of-thumb recipes have been replaced by precise formulas, uniform procedures, and repeated tests of the quality of finished products.

Eye shadow preparations illustrate the role of chemistry in today's cosmetic production. Because they are used so close to the eyes, the required pigments must be as harmless as possible. Carbon black, ultramarine, and the various yellows, browns, and reds of iron oxides provide a range of colors. Metallic lustres are made by adding finely ground

$FCCl_3$
trichorofluoromethane

F_2CCl_2
dichlorodifluoromethane

$N-CH=CH_2$

polyvinylpyrrolidone

$H_2C=CHOAc$
vinyl acetate

aluminum, bronze, silver, or gold. Pearl-like effects are created by adding a material made from fish scales.

The base into which the colors are ground usually consists of beeswax, mineral oil, lanolin, and petroleum jelly. Titanium dioxide, TiO_2, is added to give the desired degree of opaqueness. The finished cosmetic is used to emphasize the eyes and is frequently sold in "ensembles" containing ranges of shades in the same color.

Although Cleopatra's use of eye makeup would be judged a bit heavy by today's standards, this cosmetic is very popular. However, there is one remarkable difference in the use of the cosmetic today when compared with Cleopatra's time. Today, eye shadow and other cosmetics are produced in large quantities and at costs which allow them to be sold at prices which anyone can afford.

Hair-setting lotions, especially those that are marketed in aerosol or pressurized cans, also illustrate the role of chemistry in making products to improve appearance. Again, people have used greasy or sticky preparations to hold coiffures in place throughout much of human history. These preparations, however, were much too messy to gain wide-spread use.

The Norwegian, Erik Rotheim, in 1931 patented the spray can. It was first used to distribute hair-setting lotions in the United States in the late 1940's. By the 1970's, people were buying hair spray in pressurized cans at the rate of five or so each year for each woman and girl in the country. Worldwide markets added up to more than a billion cans each year.

The hair-spray can contains the actual lotion and a pressurized propellent. The can is closed with a valve that dispenses the lotion when pressed. The lotion and the propellent are forced from the can as a very fine mist. The propellent usually evaporates immediately leaving the lotion distributed uniformly on the hair.

The propellent is a non-flammable mixture of such materials as trichloromonofluoromethane or dichlorodifluoromethane. More will be said about these materials later. It is noteworthy at this point to realize, however, that hair sprays account for nearly half of all of the aerosol cans used each year in the U.S.

The fantastic growth of this cosmetic involved the development of an almost ideal container and an almost equally ideal lotion. To attract as wide a market as it did, the cosmetic had to meet several highly critical specifications. For example, it had to do well what it claimed to do—namely, keep the hair in place for a reasonably long time. It had to wash out easily. The lotion had to form a clear film on the hair, not become sticky by picking up moisture or become flaky after being on the hair for several hours.

Shellac was known to put a glossy, waterproof film on wood. It was natural, therefore, to use shellac for a hair-setting lotion, but this substance did not perform well. Not only was the shellac difficult to wash from the hair but it clogged the spray valve. Among those who looked for a better material was the German chemist, H. Kroper. He knew about a film-forming material—polyvinylpyrrolidone. The film formed by this material washed from the hair easily, but the film was hygroscopic—that is, it picked up moisture from the air and made the hair sticky in humid environments.

Wolfgang Linke, another German chemist, tells in the May 1974 issue of *Chemtech* how chemists kept the good and got rid of the bad properties of shellac and polyvinylpyrrolidone by putting together an almost ideal hair-setting lotion. They used vinyl acetate and vinylpyrrolidone for their raw materials. Individual molecules of these two substances were joined together in repeating units to form a giant particle or copolymer. This new copolymer retained the good properties of shellac and formed a glossy, nonhygroscopic film, but now the film washed from the hair easily. Furthermore, by adjusting the ratio of the two raw materials, the properties of the hair spray could be adjusted to meet the preferences of individual customers.

Individual manufacturers also add materials which improve the appearance of the hair, set the hair more securely, or counteract the damage caused by other hair treatments. Some of these additives are protein products, silicone oils, mink fat, and polyethyleneglycol. Dr. Linke emphasized that although a manufacturer can anticipate many of the prob-

Chemists looking for a substitute for shellac in hair spray settled on a film-forming material . . .

. . . which held the hair in place like shellac, but washed out easily. Unfortunately, it picked up moisture from the air.

Chemists later combined two substances, by joining their molecules, into an ideal hair-setting lotion retaining the holding power of shellac, washing out easily, and staying dry.

lems that are involved in making and distributing hair spray in pressurized cans, not until many people have used the product under all kinds of conditions could the manufacturer head off all problems. As the billion cans-per-year market indicates, however, hair spray is a very popular cosmetic.

Whether or not a person chooses to use eye makeup, hair spray, or any other cosmetic continues to be a matter of personal choice. Chemistry's role seems to be to make sure that products are available which do what they claim with minimum risk of unforeseen, unfavorable side effects.

Potassium permanganate for when we need a cool burn

It may seem strange to jump from ladies' dressing tables to the bad smells from cattle feedlots but this may point out how widely chemistry becomes involved in the things people do.

Burning has always been the easiest way to get rid of trash, but it often pollutes the air with foul and sometimes poisonous gases. Some trash won't burn or is mixed with nonflammable materials. Problems such as these call for a special kind of burning.

Fundamentally, burning usually involves the combining of oxygen from the air with whatever is being burned. If the material being burned is made up of carbon and hydrogen, for example, the carbon combines with oxygen to form carbon dioxide, and the hydrogen combines with oxygen to form water. Viewed from deep within the invisible world of individual atoms, at the high temperatures which exist during burning, oxygen atoms trap the outer electrons of other atoms. These other atoms become bonded to oxygen to form molecules of oxides. When large numbers of oxygen atoms are involved, enough heat is produced to cause the flames we associate with burning.

Chemists make a substance that allows materials to be "burned" under conditions which do not favor ordinary burning. This substance, potassium permanganate, $KMnO_4$, can be used to "burn" materials under widely different conditions. An example of one solution follows.

To produce enough meat to feed today's population, cattle growers feed their stock on open ranges but put them in feedlots to be fattened before going to market. As many as 30,000 head are penned up for up to 150 days. During this time, each animal eats about 25 pounds of feed and gains two or three pounds each day.

Each animal also produces about 25 pounds of excreta each day, and if this material doesn't dry and piles up too much, the odor spreads for miles. No "good housekeeping" methods are known that can keep the smells from causing neighbors to complain. Potassium permanganate provides a solution to the problem. At one feedlot, the owners built a 1000-gallon sprayer that could be pulled around by a tractor. One hundred pounds of potassium permanganate when dissolved in 1000 gallons of water covered a 5-acre feedlot. If sprayed three times a year, the problem of foul odors was fairly well solved.

Even the people who rely on this solution to the problem are puzzled somewhat by how well the job is done. Far greater quantities of waste materials are handled than can be accounted for by the amount of potassium permanganate used. This suggests that the potassium permanganate breaks down to form substances which catalyze the oxidation of the waste materials. On this basis, because the catalyst particles can be used over and over again, their role is to speed up the reactions that naturally convert manure to products that don't smell bad.

Potassium permanganate solutions are used in scrubbers to get rid of many kinds of smelly waste gases. Sewage plants are kept almost odor free when sewage gases are collected and bubbled through permanganate solutions. The method also works for breweries, meatpacking and rendering plants, pulp and paper mills, tanneries, and other plants that can pollute the air with smelly gases.

These solutions can also be used to "burn" impurities out of water supplies. For example, soluble iron is changed to insoluble products. At the same time, the permanganate becomes insoluble manganese dioxide, MnO_2. After settling and filtering, the water is freed from these and other impurities.

High temperatures can cause molecules to overcome the potential energy barrier and break apart when they collide. Catalysts can produce a sequence of "activated complexes" which allow the same reaction to proceed at lower temperatures.

The usefulness of permanganate as a "cool burn" can be traced to the make-up of manganese atoms. Apparently, when manganese atoms are combined with oxygen in permanganate particles, they attract electrons from whatever other atoms they bump into. When these atoms lose electrons, they are oxidized in very much the same way burning results in the formation of oxides.

The starting materials for making potassium permanganate are the ore, pyrolusite, MnO_2, and potassum hydroxide, KOH. The ore contains about 35 percent manganese. Clever and precisely controlled chemistry is needed to take the raw material particles apart and put the pieces back together in exactly the right way to make potassium permanganate. One of the most difficult problems is to combine each manganese atom with the four oxygen atoms in the permanganate ion in such a way that each permanganate ion teams up with only one potassium ion. The solution to this problem features using a flow of electrons through a solution of MnO_4 particles, and if the electron flow is controlled at exactly the proper rate, potassium permanganate crystals form and settle to the bottom where they can be harvested, dried, and marketed.

All in all, how chemists change one of nature's ores into a substance that gets rid of bad smells, purifies water, and solves other problems by controlled burning provides a good example of what chemists do and what the results are.

Some kinds of molecules are good executives or supervisors

Some kinds of chemical reactions are favored if a special kind of additional particle is present—a particle that influences the reaction but comes through unchanged insofar as its own make-up is concerned. As indicated in the previous story, such particles are called catalysts. Catalysts play such important roles in so many industrial processes that the term is used often in the broader sense of speeding up or increasing the effectiveness of all kinds of actions.

Many kinds of particles can act as catalysts, and some catalysts can influence several different reactions. Some catalysts are rather simple molecules—manganese dioxide, for example; others are complex. The enzyme, catalase, which influences the reactions whereby energy is released by the oxidation of food in our bodies, calls for thousands of atoms in each molecule.

Several metals can catalyze important reactions. When nitric acid, HNO_3, is made by the oxidation of ammonia, NH_3, a platinum-rhodium alloy plays an important role. About 2 million pounds of nitric acid are produced per pound of catalyst, and the metal can be trapped in filters and recycled.

How do catalysts work? One theory says that the catalyst's surface traps the reacting particles and thereby increases the likelihood that collisions will take apart the raw material molecules and allow product molecules to come together. By stopping particles, their motion releases energy which also may favor the formation of new kinds of molecules. In some cases, the catalyst helps to form an intermediate particle and then drops out after other steps take the newly forming molecule further along toward its final product.

Chemists make the materials needed to make the things we need and to answer "what kind" and "how much" questions

The work of many chemists is devoted to finding or creating materials that help to identify the kinds of elements or compounds present both in the raw materials with which chemists work and the finished products which solve our problems and meet our needs. It is equally important to know the amounts of each element or compound. In fact, analytical chemistry, both qualitative and quantitative, is an essential supporting service to the total profession and industry of chemistry.

Qualitative analysis can be as simple as adding a drop of vinegar to a

Much of the effectiveness of catalysts comes from their being able to cause repeated reactions without having their own molecules broken apart.

strip of litmus paper to "prove" that vinegar is acidic, or it can be as complex as using extremely sophisticated "black boxes" to determine the kinds of elements in rocks brought back from the moon. The important idea is to create analytical procedures which yield accurate and easily interpreted data.

One of the newer solutions to analysis problems that illustrates nicely the tactics and strategy of this branch of chemistry is a test strip which allows a person to identify significant changes in the composition of urine. Each strip has seven separate reagent areas for testing the acid-base properties of urine and for detecting the presence of proteins, glucose, ketones, bilirubin, blood, and urobilinogen. Test results may provide information regarding carbohydrate metabolism, kidney and liver function, and acid-base balance.

Each of the reagent areas contains a combination of carefully selected molecules that will produce unique colors when the corresponding urine component is present. Furthermore, the intensity of the color reflects the amount of each component. On this basis, the results obtained by using one of the test strips can be "read" simply by comparing the color of each reagent area with a set of color standards.

Obviously, there must be qualitative and quantitative analytical procedures for all of the materials with which chemists work. This calls for thousands of different kinds of reagents and test procedures as well as highly sensitive instruments. Because each kind of molecule has unique properties, however, chemists know that these unique properties can be taken advantage of to solve "what kind" and "how much" problems.

Plastics

Museums throughout the world are filled with proof that people have always managed to find materials with the properties they needed to fashion or build the things their minds and hands created. The use of stone and clay, wood and bone, horn and ivory goes back many years. Gold and silver, iron, copper, and other metals, including such alloys as brass and bronze, have been used to make the things people needed for thousands of years.

The 1900's opened a new chapter in the history of our search for materials, a chapter that can well be called the Age of Plastics. In 1909, Leo Baekeland found that the phenolformaldehyde that Adolph Baeyer had made 30 years earlier could be molded into articles and dissolved in solvents to make varnish. With this discovery, a pathway was opened that has led to all kinds of new materials—materials that people can use to make all manner of things.

In many ways, we are still growing up with the Age of Plastics. Bakelite and cellophane, Plexiglas and masonite, fiberglass and vinyl, polyethylene and polyester—these are all relatively new terms, but they have already become "household" words. They will no doubt be joined by even newer names because new plastics are being created every day.

The Age of Plastics began with the rather random puttings together of materials at hand. However, just as students are puzzled when they see a pink solid gush from a tall cylinder when hydrogen chloride gas is bubbled through a mixture of phenol and formaldehyde, the early discoveries in the field of plastics led promptly to "how" and "why" questions.

These questions led to the idea of polymerization. Rather than taking apart the raw material molecules and using the pieces to put together new molecules, making plastics is more likely to involve the joining of many of the same small molecules to produce a giant molecule. With the change from a single particle—the monomer—to a giant particle—the polymer—comes spectacular changes in properties. Colorless liquids or gases, for example, become solids, foams, films, or fibers. Some polymers are colored, some are colorless.

Nature provides many examples of polymerization. Single molecules of glucose and fructose, for example, can come together, fragments can break off to become water molecules, and sucrose or table sugar will be formed. Glucose molecules polymerize to yield starch. Longer chains of glucose molecules give cellulose.

Bakelite is a polymer made by the polymerization of phenol and

PhOH
phenol

$H_2C{=}O$
formaldehyde

$H_2C{:}CH_2$
ethylene

$H_2C{:}CHPh$
styrene

$H_2C{:}CHMe$
propylene

$H_2C{:}CMeCO_2{}^-$
methacrylate

MeOH
methanol

$H_2C{:}CMeCO_2H$
methacrylic acid

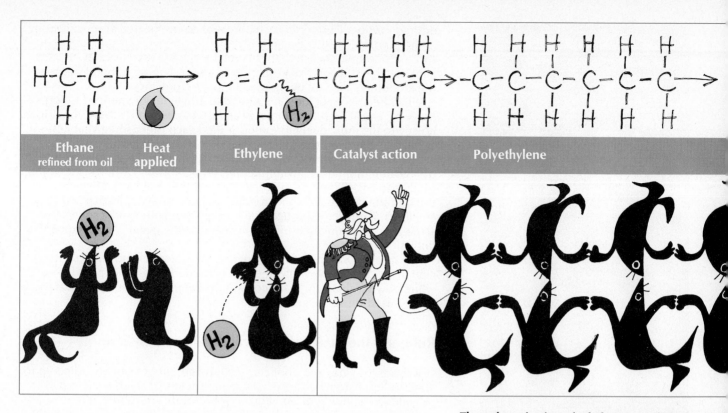

| Ethane | Heat | Ethylene | Catalyst action | Polyethylene |
| refined from oil | applied | | | |

The polymerization of ethylene is a relatively simple example of using catalysts to cause small molecules to join in the precise pattern required to yield highly useful giant molecules or polymers.

formaldehyde molecules. If hydrogen chloride molecules are on the scene when these small monomers come together, water molecules are produced from fragments of the monomer molecules. Where these fragments left the monomer molecules, the "broken ends" join to form the polymer. This can happen "poly" times and thus yield a giant Bakelite molecule. The Bakelite can be mixed with wood flour or other fillers and heated in molds under pressure to make plastic items or structural materials.

Another step in the Age of Plastics occurred in the 1930s. Methyl methacrylate was found to polymerize to yield a highly adaptable, clear plastic. Methacrylate is made by the reaction of methanol with methacrylic acid, and both of these raw materials are obtained as petrochemicals—that is, by modifying the molecules that are in petroleum.

After the discovery of the acrylic plastics, Plexiglas, Lucite, Perspex, and similar products became very popular. Today, however, an even greater number of uses are found for polyethylene, polystyrene, polypropylene, and similar members of a new family of plastics.

The story of this group of plastics features the work of Karl Ziegler of Germany and Giulio Natta of Italy. These men received a Nobel Award in 1963 for devising a way to get monomers to polymerize in a precise pattern. Dr. Ziegler acknowledged the many practical applications of this group of plastics when he gave his Nobel Award lecture. He looked upon the popularity of these plastics as an explosion and gave credit to the many chemists, engineers, and designers whose ingenuity and creative imaginations were responsible for using these plastics in so many ways.

Using polyethylene as an example, the usefulness of this group of plastics depends very much on having the individual monomer particles join in exactly the proper way. Not only must each monomer hook on to the proper spot on another monomer but they all must be "facing" the proper direction. This is where the work of Ziegler and Natta became so important.

They found ways to attach metal atoms to ethylene and other kinds of hydrocarbon molecules by replacing one or more of the hydrocarbon molecule's hydrogen atoms. These new compounds were called metal alkyls. They are almost explosively reactive, an advantage as well as a disadvantage. If they can be controlled, these molecules can carry large bundles of energy to the site of a chemical reaction at the same time they bring fragments of molecules that are needed to put together the

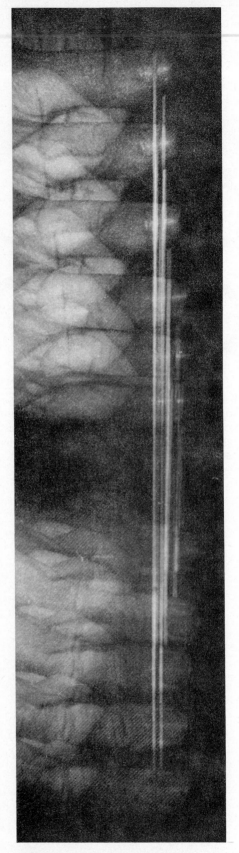

Rubber snaps back because forces within and between trillions of molecules can be stretched . . .

desired product. Furthermore, with metal atoms located at specific points in their molecules, it is apparently easier to control their orientation in space. Consequently, when these metal alkyls are used as catalysts, each kind of catalyst favors the formation of a specific kind of bond between the monomers that are being used to make a polymer.

It had been known for some time that ethylene polymerized to form polyethylene at high temperatures and pressures. Under these conditions, however, the individual monomers joined each other somewhat randomly. Ziegler used one of his "organometallic mixed catalysts" to polymerize ethylene at lower temperatures and pressures. The catalyst was triethylaluminum mixed with titanium tetrachloride, $TiCl_4$. He added the catalyst to a gasoline-like hydrocarbon and then bubbled ethylene through the mixture. Within an hour he had bubbled between 300 and 400 liters of ethylene through one liter of the mixture and a solid substance had accumulated sufficiently to cause the material to become too difficult to stir.

The catalyst was then destroyed by adding alcohol and air. The precipitate became snow white and was filtered off as a white powder—a white powder that can be fashioned into the polyethylene products so widely used these days.

Rubber—the material that will do almost anything

Rubber stretches and bounces. If stretched, rubber can be pulled up to 10 times its unstretched length. When it is let go, it snaps back to its original size and shape. Similarly, squeeze or distort rubber, and it springs back to its original shape. Rubber is also waterproof and airtight. It doesn't wear away easily; in fact, rubber withstands friction and wear better than steel. It can be bonded almost permanently to fabrics, metals, glass, or plastics. It is unchanged by contact with nearly all other materials.

Rubber does so many things that between two and three pounds are used by each person each year. Here in the United States, if our only source of rubber were from rubber trees, we would need a rubber plantation the size of Ohio and Massachusetts combined.

The Indians of South and Central America were first to recognize and take advantage of the properties of rubber. Spanish explorers found the Indians playing a game with a ball that bounced better than anything known in Europe. In time, the Indians learned to spread rubber on clothes to make them waterproof and to shape rubber into waterproof shoes and unbreakable bottles.

Until rubber became available, there was no equally flexible or waterproof material. Pig bladders were used for inflated balls or balloons. Firefighters used leather hose which leaked at every seam. Leather boots were never really watertight. Even Joseph Priestly had to use bread crumbs to rub out pencil marks until he obtained the newer rubber. In 1803, the first rubber factory was established in Paris. It made elastic bands for garters and braces. Other inventive people used rubber to "rubberize" the silk that was used to build pioneer balloons.

The rubber of the early 1800s had shortcomings. It was sticky, and in hot weather it got soft and sticky. In cold weather it became so stiff it was almost inflexible. In 1823 a Scotsman, Charles Macintosh, solved some of these problems. He sandwiched the sticky rubber between two layers of closely woven cotton.

In 1839 Charles Goodyear discovered a much better solution to natural rubber's shortcomings. He found that when rubber was heated with sulfur, it became the almost ideal material we know as rubber today. The process is called vulcanization. Vulcanized rubber found so many uses that by the end of the 1800s there was a worldwide rubber shortage. As a result, new rubber plantations were established in other parts of the world, in some cases using rubber tree seeds smuggled out of South America.

This didn't solve the rubber shortage immediately, and by 1910 the price of rubber was more than three dollars per pound. This high cost prompted the chemists of the day to look into the possibilities of producing synthetic rubber. As early as 1826 Michael Faraday found that natural rubber's molecules consisted of five carbon atoms for every eight

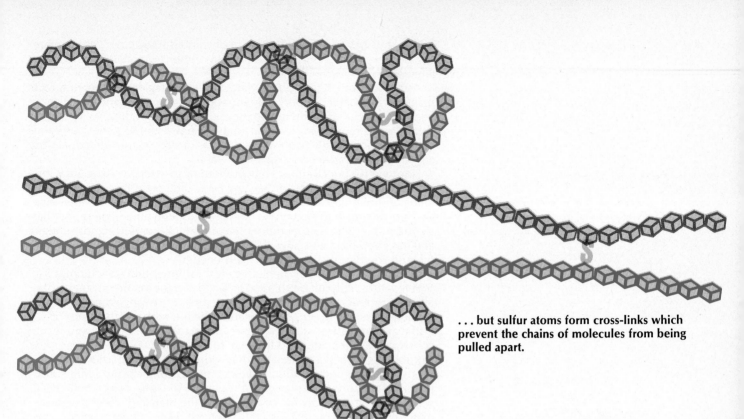

. . . but sulfur atoms form cross-links which prevent the chains of molecules from being pulled apart.

hydrogen atoms. In 1860 G. Williams heated rubber in a closed container and collected isoprene, a liquid that boils slightly above room temperature and whose molecules have five carbon and eight hydrogen atoms.

For years chemists tried to make rubber by combining isoprene molecules. Then came World War I. Germany was cut off from supplies of natural rubber and consequently the German government encouraged chemists to develop synthetic rubber. By 1916 Germany was producing 150 tons of synthetic rubber each month from dimethylbutadiene, a compound similar to isoprene.

In the middle 1920s Hermann Freiburg made his Nobel Award winning discovery that rubber is made up of giant molecules. In today's language isoprene is the monomer of the polymer called rubber. Even though the price of natural rubber dropped as low as five cents per pound, the development of synthetic rubber continued full speed in Germany and in Russia. By 1940 production of synthetic rubber reached 175,000 tons in Germany and 90,000 tons in Russia.

In 1942, when Japan captured all of the major sources of natural rubber in Asia, the U.S. launched a vast effort to produce synthetic rubber. Before World War II was over, production had been pushed from 10,000 to 700,000 tons per year. It is interesting to see how political and economic circumstances can influence and be influenced by efforts to understand how rubber molecules are made and how to take over their production.

As is true with all substances, the unique properties of rubber are a result of the kinds and arrangement of atoms in its molecules. Its elasticity is the result of several thousand isoprene molecules' being joined in a chain. The permanent bounce and elasticity of vulcanized rubber come from cross-links between chains, cross-links which allow the chains to be stretched but not separated.

To put together the kind of rubber molecule a chemist has in mind, however, calls for know-how and at least a bit of good luck. The rubber may or may not have the properties the chemist hopes for, and know-how is required to get the monomer molecules to join in the proper way. Trillions of isoprene molecules must be linked in precise patterns, for example, to make no more rubber than is needed for a rubber band. To

$$H_2C = CHCMe = CH_2$$

isoprene

$$H_2C = CMe\ CMe = CH_2$$

dimethylbutadiene

add to the challenge, if molecules of unwanted substances contaminate the isoprene during polymerization, the rubber's characteristics can be affected.

As pointed out earlier Giulio Natta and Karl Ziegler discovered how to make catalysts that not only help monomers to join together but also keep them lined up in definite patterns. Chemists continue to hunt for catalysts and monomers that will enable them to put together new kinds of rubber. There have been so many kinds of rubber created that almost any kind of rubber seems possible.

Isoprene rubber combines many of the properties we associate with rubber. One or more properties, however, can be emphasized in other kinds of rubber. Styrene-butadiene rubber has good abrasion resistance and less tendency to skid on wet or icy roads. Polybutadiene rubber has more bounce than natural rubber. Neoprene rubber withstands sunlight and has good heat resistance. Nitrile is the rubber to use when a tough, oil- and chemical-resistant material is needed.

Since butyl rubber is airtight, it is ideal for inner tubes and linings for tubeless tires. Polysulfide rubber is a good sealant for aircraft fuel tanks. Silicone rubbers stand up under wide temperature ranges and are extremely inert. These properties enable silicone rubber to be used for artificial heart valves, stoppers for drug and medicine bottles, and for conveyor belts that are used to move foods.

Rubber is indeed a material with a wide range of properties. These properties have been used to solve countless problems and fulfill countless needs. Future contributions of rubber to meeting our material needs seem to be limited only by the abilities of the men and women who match their minds and hands against what it takes to find new monomers and to control their polymerization.

Special kinds of molecules make the clothes we want and need

Before about 1900 people depended upon plants and animals for their clothing. We can only guess how they learned to twist hairs or plait fibers together to form thread or yarn. Also, who knows when people first became curious about how wool grows on a sheep's back, how silky filaments are spun by silkworms, how the long fibers in the stems of flax plants are produced, or how the fluffy white bolls develop around the seeds of cotton plants?

In 1855 George Audemars treated mulberry leaves, the favorite food of silkworms, with nitric acid, HNO_3, and produced nitrocellulose from which he obtained crude fibers. Thirty-five years later Count Hilaire de Charbonnet improved this process by dissolving the nitrocellulose in ether and alcohol. In 1857 Schweitzer discovered that cellulose, the chief molecule in cotton, could be dissolved in a solvent prepared by dissolving copper hydroxide, $Cu(OH)_2$, in ammonia water. Thirty-three years later Louis Henri Despaissis obtained a patent on a process using this solution to dissolve cellulose and then spin this solution into a kind of textile fiber we now know as rayon.

A better way to make rayon was developed in 1893. Cellulose from wood pulp or from cotton fibers that are too short to make into thread were treated with a strong solution of sodium hydroxide, NaOH, followed by carbon disulfide, CS_2. The cellulose formed a thick, viscous syrup. In fact, this method for making rayon was called the viscose process. When this syrup was forced through tiny holes in a metal spinneret, hairlike fibers were produced. Sodium hydroxide and carbon disulfide were removed from the fibers by sending them through a water bath that contained sulfuric acid, H_2SO_4. Now the cellulose molecules that were originally joined together in the wood or cotton were rearranged. They were now more orderly and, in some respects, made fibers which could be woven into a better cloth.

During the 1900's several companies produced rayon in larger and larger quantities. Many improvements were made. For example, spinnerets were made from metal that resisted corrosion. As many as several thousand holes one-thousandth inch in diameter in each spinneret produced rayon in the form of a thick rope of fibers or tow. The name,

$H_2N(CH_2)_6NH_2$

hexamethylene diamine

$HO_2C(CH_2)_4CO_2H$

adipic acid

$(\ —NHCO(CH_2)_5— \ )_n$

nylon

68

rayon, means all of the brilliance of a ray of sunlight, and dyes were developed that enabled rayon to live up to its name.

By 1911 8 plants produced 18.7 million pounds of rayon each year. By 1940 production had grown to 1.96 billion pounds annually and to more than 6.5 billions by 1966. Production increased by 2 billion pounds each year, and by 1968 rayon was claimed to be the world's most versatile fiber. Ways had been found to create rayon that was water repellant, flame resistant, and shrink and crease resistant.

It is interesting to compare rayon with natural cotton fibers. In natural cotton fibers the individual cellulose molecules are joined by natural forces. In rayon, the cellulose molecules have been separated and then rejoined in a way that favors a specific arrangement. The differences between cotton and rayon result more from differences in the arrangement than in the kinds of molecules in the fibers. In cotton or wood pulp as many as 2000 cellulose molecules form the polymer. Rayon fibers consist of 350–450 cellulose molecules per fiber.

The differences between cotton and rayon are a good example of taking things apart and putting things together to meet today's needs. However, rayon was only the first of a new textile fiber synthesized during recent years.

In 1935 a group of chemists led by Wallace Carothers was studying polymerization—a process of joining molecules. One of their polymers consisted of the 2 monomers, hexamethylene diamine and adipic acid. When this polymer was forced through a hypodermic syringe, a thin, hairlike stream came from the needle, cooled, and solidified. Nylon had just been invented. At first, nylon was used more as a plastic than as a textile fiber. One of its first uses was to replace hog bristles in toothbrushes, but when nylon stockings were marketed in 1939, this new textile fiber skyrocketed to popularity.

Other new textile fibers were quickly created—Terylene, Dynel, Orlon, Dacron, Spandex, Herculan, Acrilan, Kodel, and other variations of more than a dozen textile fibers. Rayons were joined by acetates and triacetates, nylon and other polyamides by polyesters, acrylics and modacrylics, and the olefin yarns, polypropylene and polyethylene.

Paralleling the development of new textile fibers has been the search for ways to treat textiles to make them more comfortable, attractive, and easy to care for. Today 45 pounds of textile fibers are produced each year for each person in the United States. Fifty years ago this figure was 30 pounds, and 100 years ago it was only about 13 pounds.

The highly flame-resistant clothing which protects many fire fighters is a good example of the wide range of properties that can be built into modern textile fibers.

Phosphates, materials which play many roles

In 1776 phosphorus was known mainly as an element that glowed in the dark. Today, phosphorus, especially phosphates, plays many roles in helping us live well and enjoy life.

Phosphorus compounds are a vital constituent in all plant and animal cells. They pass energy between molecules which need energy and molecules which have stored energy. In a different role but still within living systems, phosphates help to form bones. We will see later that phosphates also play many roles in the non-living world.

Before 1900 one of the main sources of phosphorus was animal bones such as those from the stockyards and packing plants of Chicago. Hennig Brandt in Germany discovered elemental phosphorus accidentally in 1669 while he was working with urine to create the philosopher's stone. He didn't achieve eternal wealth, but he did exploit the commercial properties of this element that glows in the dark—an easy task in those superstitious times.

Phosphate ions join hydrogen ions to form phosphoric acid, H_3PO_4. In turn, phosphoric acid in water yields hydrogen ions and phosphate ions, PO_4^{-3}. All acids will give up hydrogen ions; in fact, the usefulness of acids depends on their making hydrogen ions available. Stomach juices, for example, digest food partly because the juices contain hydrochloric acid, HCl.

Phosphoric acid is an adaptable acid. Each phosphate ion ties up three hydrogen ions but releases them one at a time. The actual number released depends on the conditions and other kinds of molecules or ions present. Different quantities of energy are involved when one rather than two or three hydrogen ions is given up.

Phosphoric acid is used to clean metals and prepare their surfaces for painting or other special treatment. Much aluminum owes its brightness to this acid. Steel is often cleaned and rust-proofed by being dipped in a bath containing phosphoric acid and small amounts of zinc or manganese. A thin coating of metal phosphates is formed on the iron and protects it from corrosion.

The role of phosphates in soaps and detergents has been publicized widely, not because of their importance in these products but because of their effects when left in waste wash water. These effects are related to the use of phosphates in fertilizers. Phosphorus is one of the chief building blocks needed for plant growth. How well plants grow in many environments depends on the amount of phosphates available. One aspect of water pollution involves the excessive growth of algae and other aquatic plants when excessive concentrations of phosphates build up in bodies of water.

Phosphate rock deposits in Florida, North Carolina, and several other states provide the raw material for phosphorus and its compounds. The rock is mined, ground, and mixed with coke and sand. The mixture is then heated to 1550°C. At this temperature, the carbon atoms in the coke pull the other atoms away from the phosphorus in the rock, and pure phosphorus boils from the mixture. Solid phosphorus forms when the vapor cools.

All of this must be done in airtight systems. Phosphorus is extremely combustible, and when exposed to air, it ignites spontaneously. Consequently, it is stored and shipped under water. Before it is used, however, most phosphorus is converted to phosphoric acid by burning the phosphorus in air and dissolving the oxide in water.

Phosphoric acid can also be made by treating phosphate rock with sulfuric acid. Since this is a cheaper method, it is displacing earlier ones. After phosphoric acid is made, its acid property can be adjusted. If it is to be used to add tartness to soft drinks, for example, its acidic properties are played down. This is done by allowing the acid to react with washing soda, Na_2CO_3. A sodium ion replaces one of the hydrogen ions in the phosphoric acid. In addition to its use in soft drinks, the resulting monosodium phosphate, NaH_2PO_4, is used in puddings and baby foods or in effervescent laxatives and preparations for upset stomachs. Because monosodium phosphate has a mildly pleasant taste, it can be used whenever the acidity of foods must be increased.

Phosphoric acid can be used to make sodium pyrophosphate, $Na_4P_2O_7$. This compound is used in preparing large batches of dough or batter. Since the pyrophosphate reacts slowly with baking soda, the dough or batter can be made up commercially and baked at home when it is convenient.

Pyrophosphate particles tend to trap other kinds of atoms, especially iron, calcium, or magnesium. Sometimes potatoes discolor while being cooked because of their iron content. If some pyrophosphate is added to the cooking water, the potatoes remain white. When added to dish water, pyrophosphate traps (sequesters) any calcium or magnesium in the water, thus softening the water and preventing scum or film from forming on the dishes.

The two hydrogen ions in monosodium phosphate can also be replaced by sodium ions. The resulting compound stabilizes evaporated milk and cheeses for storage. If trisodium phosphate, Na_3PO_4, is added to the brine in which hams are cured, the hams are tenderer and juicier when cooked and smoked. Starch that is treated with this compound forms gels in cold water for instant puddings.

Trisodium phosphate can be combined with abrasives and bleaches to make scouring powder. It has long been used to soften water because it precipitates calcium and magnesium ions. Calcium phosphate, $Ca_3(PO_4)_2$, itself is a useful compound in processing foods. It absorbs other substances, particularly water. Dry foods that tend to get sticky by absorbing moisture from the air are protected when a little calcium phosphate is included. This compound not only steals any moisture that is available but through its calcium content adds to the nutritive value of the food.

In 1776 phosphorus was only a strange element. By 1876 its properties were becoming known well enough to allow its use in making matches. A few of its compounds were known and used—baking powder for example. By 1976 millions of tons of phosphorus and its compounds are being used to solve numerous problems and to fulfill many needs.

New applications are being discovered all the time. When it was learned that sodium aluminum acid phosphates could be used to help prepared cooking mixes retain their ability to rise during baking no matter how long they were left on the shelf, another large market was created. Biscuits, cakes, and waffles made from such prepared mixes compete quite well with those made "from scratch."

Research sometimes puts new chemicals "in the bank"

Many people devote their lives to making new materials, and their motivations are not simply the desire to solve immediate problems or fulfill immediate needs. If a discovery turns out to be useful, their sense of achievement may be heightened. Similarly, when practical chemists make a discovery—and there seems to be at least a bit of the research spirit in most chemists—they are proud to see their discovery become

From Faraday's sketch of the first dynamo

Leo Hendrik Baekeland
1863-1944

Beginning with receiving $1,000,000 for the invention of a better photographic printing paper, Leo Baekeland became one of the chemical industry's most successful inventors and businessmen. His Bakelite created a milestone in the plastics industry.

Michael Faraday
1791-1867

Best known for his discoveries in electricity, Michael Faraday was very much an "all-around" scientist. He spent many years investigating the oily byproduct from the artificial illuminating gas industry. These investigations led to the isolation of many new compounds.

Wallace Hume Carothers
1896-1937

Dr. Carothers began his career in chemistry as a college professor. By moving his research to a large chemical industry, he gained better facilities than were available in most university laboratories. Nylon was a direct outgrowth of his fundamental research on polymerization.

part of the knowledge and knowhow of chemistry.

Sometimes a substance has been known for a long time, and its properties are familiar. Ingenuity enables researchers to take advantage of this knowledge to solve problems or fulfill needs. This point is illustrated by the story of "Lake Stinking Water."

Lake Winnebago is the only convenient source of water for several Wisconsin cities. A legend explains the name of the lake as a souvenir of the Indians' description of the foul odors that come from the lake especially in late summer.

The streams which feed Lake Winnebago drain thousands of acres of Wisconsin farmland. Farmers and dairymen use fertilizer on this farmland to provide nitrates, phosphates, and potash for their crops. The wastes from livestock produce manure which is rich in plant nutrients. The combination of abundant nutrients in the lake and abundant sunlight during late summer yields an environment that is ideal for plant growth, especially algae. But growth leads to maturity and, eventually, death. With death comes decay. With decay comes the taking apart of once living material and the putting together of simpler molecules. Sometimes these decay products have very unpleasant odors, unpleasant enough to give a lake the name, "Lake Stinking Water."

Local people tend to take Lake Winnebago's odor for granted, but a bad situation turned worse in late August 1939. The summer sun, high in the northern sky, was radiating maximum sunlight into the lake. Run-off from farm fields was adding plant nutrients. Conditions were ideal for an algae "population explosion." Algae counts rose to millions per glassful. Odor and turbidity were so bad that municipal water treatment plants couldn't possibly produce clear, odorless water.

The cities which drew water from Lake Winnebago joined forces. They tried every type of water treatment system on the market. None was adequate. Finally, a solution to the problem shaped up in the minds of Andrew J. Marx and his co-workers at the Menasha water treatment plant.

These people knew that copper sulfate, $CuSO_4$, prevented growth of algae in aquariums and swimming pools. They also knew that it would be totally impractical to try to build up the high concentrations of copper sulfate needed to kill the algae in Lake Winnebago. The cost would have been prohibitive, and the effect on the environment of the lake was unpredictable.

Then came the idea to build a small reservoir, fill this reservoir with Lake Winnebago water, but treat the water with copper sulfate while it was flowing from the lake to the reservoir. This way the algae would be killed before they scattered through the whole reservoir. To test this idea, Marx and his fellow engineer-scientists treated batches of water for several days with copper sulfate and let the treated water stand in open tanks. Daily records were kept of water conditions, changes in odor and turbidity, and the algae count. Results were amazingly favorable.

The key to the success of the project was the realization that the length of time the water had to stand depended on the weather. On warm sunny days three days was long enough, but seven days were needed on cool cloudy days. The project succeeded because it used the cause of the problem to produce the cure. The environmental conditions which accelerated algal growth also hastened its decay. This told Marx and his team that they could add the same amount of copper sulfate each day to the intake water, and natural environmental conditions would automatically adjust to the amount of algae in the lake.

At this point, Menasha's troubles were turned over to land surveyors and construction engineers. A suitable location was found for a full-scale reservoir with a channel that let Lake Winnebago water flow through a pipe under the chemical feed house and then into the reservoir. The story of "Lake Stinking Water" ends on a successful note because scientists, engineers, and civic officials joined forces to solve an environmental problem.

This is also an appropriate note on which to end our stories about what chemists make for people. Chemistry provides the wherewithal to solve our problems and fulfill our needs but solutions to problems and fulfillment of needs also depend upon inventive and creative thinking on the part of scientists, engineers, and civic officials.

5

Amplifying Our Power: The Chemistry of Energy

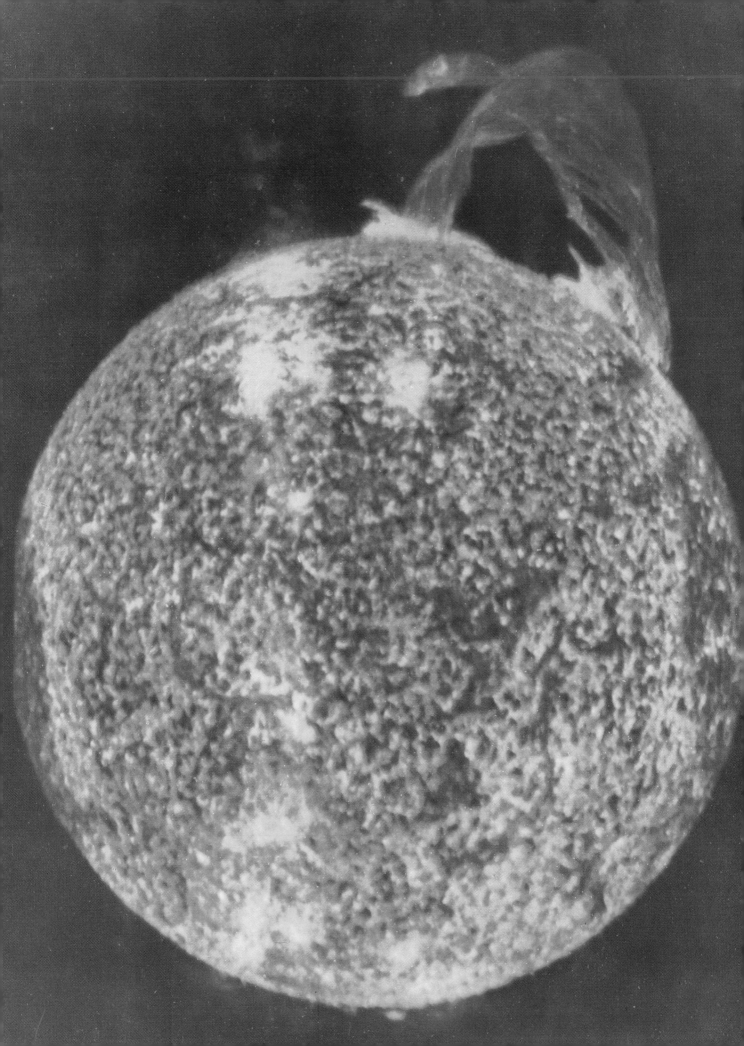

Nothing lives or moves without expending energy. It is as though everything that happens is the action of an enormous, fantastically complex wind-up toy. If the toy were to run down, if energy were no longer available, the world would become motionless, cold and dead.

The relentless flow of energy through the universe "powers" the machines that do our work. Energy in food powers the muscles of the animal world and drives all life processes. This is a part of the stories that are told in Chapter 2, INSIDE US: THE CHEMISTRY OF STAYING WELL, and in Chapter 3, SUPPORTING US: THE CHEMISTRY OF FARMING.

The unique effects of energy on certain kinds of materials produce light, sound, and radio and television waves—part of the story told in Chapter 6, EXTENDING OUR SENSES: SEEING FURTHER WITH CHEMISTRY.

In many ways, the advance of civilization is marked by our overcoming the limitations of bare hands and muscle power. People learned early in human history to use machines and they were not long in discovering that energy is released when fuels burn. Much more recent was the discovery that energy becomes available when atoms are taken apart or put together.

Buried sunshine provides twentieth century energy

Usually plants absorb the sun's energy, grow, then die and decay. The energy built into their body tissues either escapes to continue on its relentless path to wherever energy disappears, or it is temporarily transferred to the body of some other living system.

Some 500 million years ago a set of highly fortunate circumstances occurred. Lush plant growth was buried in fine clayish sediments along seacoasts, particularly near river deltas. The plant debris was protected from decay, and thick layers of deeply buried plant material became huge chemical factories. Roots, stems, and leaves were taken apart, perhaps with the help of microorganisms, and their building blocks put back together as coal, oil, and gas. Most importantly, these substances retained much energy in their molecules.

These are today's oil, coal, and gas deposits. Prudhoe Bay, a recently discovered oil field in Alaska, is an example. The oil bearing strata are very porous sandstone laid down some 200 million years ago. The sandstone layer is up to 600 feet thick and about 8500 feet below the surface. It covers an area about 45 by 18 miles.

The Prudhoe Bay field is believed to hold more than nine billion 42-gallon barrels of recoverable oil and 26 trillion cubic feet of gas. The oil and gas have been trapped in the slightly sloping layer of sandstone because it butts up against another rock layer that is impervious.

The Ghawar field in the Persian Gulf of Saudi Arabia is about 900 square miles in area and holds an estimated 75 billion barrels of oil. The oil is in a rock formation that seems to have been laid down as limey sediment that collected in a sea or large lake. The oil has been trapped

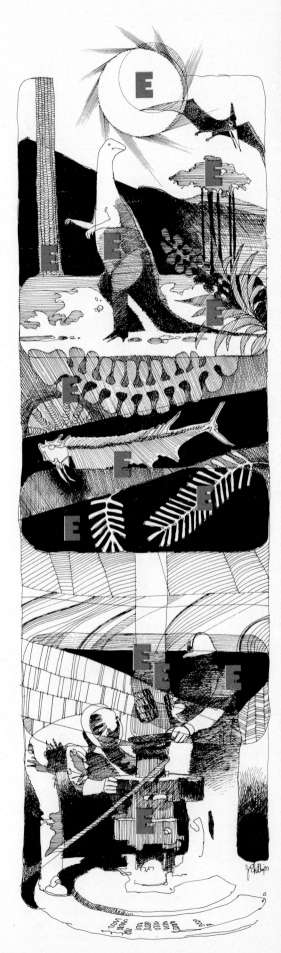

Today's gas, oil, and coal are the products of natural-energy-storing chemical reactions that took place millions of years ago.

underground for millions of years because a dome-shaped layer of non-porous rock acts as a seal above the porous carbonate rocks which hold the oil.

Before the middle 1800s, crude oil was a black, sticky liquid that seeped out of rocks although it was sometimes used in homemade medicines. In 1859 Edwin L. Drake drilled a 70-foot well and struck oil. Drake also helped create a market for oil by learning how to distill kerosene from the oil. Kerosene was a cheaper and better fuel than whale oil. Other wells were drilled to make Titusville, Pa., the site of the first U.S. oil field.

In the early 1900s the increasing popularity of the automobile created a growing demand for gasoline and of course crude oil. It is thought-provoking, however, to learn that as much oil was burned during the 10 years between 1957 and 1966 as during the first 100 years of the oil industry, that is, between 1857 and 1956.

The composition of crude oil differs widely from field to field. It averages around 85 percent carbon and 15 percent hydrogen with small amounts of sulfur, nitrogen, oxygen, and at least traces of nearly all of the elements found in seawater. Crude oil usually contains traces of porphyrins, compounds that are formed by the breakdown of chlorophyll and the hemin portion of hemoglobin. Since these molecules are hallmarks of the plant and animal worlds, their presence in crude oil suggests how oil was formed.

Crude oil is a mixture of hydrocarbons, molecules of carbon and hydrogen atoms only. Methane is the hydrocarbon with one carbon and four hydrogen atoms per molecule, but it is so volatile that it evaporates from crude oil to become natural gas. Ethane, the hydrocarbon with two carbon and six hydrogen atoms per molecule is less volatile than methane. The decrease in volatility continues with propane, the three-carbon atom hydrocarbon, and with butane, the four-carbon atom hydrocarbon. Propane and butane are easily boiled out of crude oil and are bottled under pressure to be used as bottled-gas fuel.

Hydrocarbons with five to 11 carbon atoms per molecule may be suitable for gasoline, depending upon how the carbon atoms are arranged in the molecule. The carbon atoms may be joined in a continuous chain to form normal hydrocarbons or in branched chains to form isomers. Normal heptane, with seven carbon atoms per molecule in an unbranched chain, causes engines to knock. Isooctane, with eight carbon atoms per molecule but with the carbon atoms arranged in a molecule with three branches, makes very good antiknock gasoline.

A mixture of normal heptane and isooctane can be used to standardize the suitability of hydrocarbons for gasoline. For example, a 75-octane gasoline is a mixture of hydrocarbons that makes as good gasoline as a mixture that is 75 percent isooctane and 25 percent normal heptane.

High octane gasolines consist of hydrocarbons that usually have no more than five carbon atoms in a continuous chain. It is difficult to explain why one arrangement of atoms causes an engine to knock whereas a different arrangement of the same number of atoms makes an isomer with a high octane rating.

Each carbon atom has four electrons that it can share (form valence bonds) with other atoms; valence bonds join atoms into molecules. Carbon atoms can be joined by sharing one, two, or three pairs of electrons—that is, by single, double, or triple valence bonds. Obviously, when carbon atoms in hydrocarbon molecules are joined by double or triple bonds, fewer bonds are available to hold hydrogen atoms than in the corresponding singly bonded hydrocarbon.

Hydrocarbons with double and single bonds between carbon atoms are called olefins and alkynes or unsaturated hydrocarbons. In general, unsaturated hydrocarbons aren't as good for gasoline as their corresponding saturated hydrocarbons. Hexane, for example, makes better gasoline than hexene in which two carbon atoms share a double bond or hexyne in which two carbon atoms share a triple bond. Also hexene contains two fewer hydrogen atoms and hexyne four fewer hydrogen atoms per molecule than hexane.

To see how this relates to gasoline, suppose that from a 42-gallon barrel of crude oil only 15 gallons of hydrocarbons can be boiled off. Of the remaining 27 gallons, some of the hydrocarbons aren't suitable for

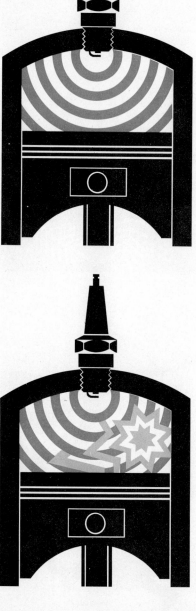

Unless long-chain molecules of hydrocarbons are converted into shorter, branched-chain molecules, they do not burn smoothly in the split-second explosions of an engine's cylinders.

gasoline because they have too few carbon atoms per molecule, others because they have too many. Still others wouldn't make good gasoline because the atoms in their molecules are arranged in unbranched rather than branched chains.

Since gasoline brings the best price of all of the products that can be made from the hydrocarbons that are most abundant in crude oil, here is a perfect challenge for the men and women whose careers involve remodeling molecules. Perhaps molecules with too few carbon atoms per molecule can be made to pair up. Perhaps molecules with too many carbon atoms can be broken or cracked in halves or thirds. If these changes create molecules with double or triple bonds, perhaps hydrogen atoms can be used to change unsaturated to saturated hydrocarbons. Finally, perhaps hydrocarbons with long, unbranched chains of carbon atoms can be taken apart and the fragments of their molecules reassembled as branched chain molecules.

We must keep economics in mind, however. Hexadecane molecules, for example, could be cracked to form two molecules of octane, but hexadecane makes good diesel fuel. Perhaps the cost of cracking hexadecane into octane would be more than the difference in the selling price between gasoline and diesel fuel. Similarly, two propane molecules might be joined to form a hexane molecule but maybe propane has as good a market price as gasoline.

Let's follow a barrel of crude oil on its way through a refinery. The crude oil is heated in a still to between 315° and 370°C, and the vapors are sent up a fractionating column that is about 150 feet high. The still is a closed vessel or pot, and the fractionating column will collect each component or fraction of the hydrocarbons in the crude oil.

Inside the fractionating column is a stair-step of 30 to 40 shallow, bowl-shaped trays which collect liquids. The rising vapors are separated, and each fraction collects in its own tray according to its boiling temperature. In other words, each tray collects hydrocarbons that condense at lower and lower temperatures. After the temperatures in the column stabilize, the top tray will be collecting the hydrocarbon in the crude oil that has the lowest boiling temperature. The hydrocarbon in the crude oil with the highest boiling temperature will be either on the bottom tray or still in the pot. In between are the trays that collect the fractions that make good gasoline. All of the trays are fitted with pipes that allow the liquids to be drawn off and sent to further refining or storage.

This is a simplified account. Engineers and technicians must make sure that temperatures are held at the proper levels, that the condensed liquids are sent back down the column, if necessary, to improve separations, and that other adjustments needed are done. Components with very low boiling temperatures may have to be taken off the top of the fractionating column as gases. Similarly, solids may have to be chopped out of the residue at the bottom of the pot.

We can't overlook the people who constantly guard against fires and explosions. In fact, as this is being written, a newscaster is announcing the recovery of bodies from a Philadelphia refinery fire. There are enormous fire hazards in handling substances which contain as much energy as is packed into the components of crude oil, and additional energy is used in many refinery processes. However, the energy stored in crude oil is not accessible without refineries.

Remodeling molecules increases yields of gasoline from crude oil

As early as 1910 people knew how to break long chain hydrocarbon molecules into shorter chains by using high temperatures and pressures. By the 1930s the cracking process was operating at lower temperatures and pressures by passing the hot, long chain hydrocarbons through clays which acted as catalysts.

As is generally true of catalysts, the clay molecules were not broken apart, and they could be cleaned or regenerated and reused. This type of catalytic cracking could convert 50 percent of the too heavy hydrocarbons to the molecules suitable for gasoline.

Today highly successful systems combine two sets of catalysts. One

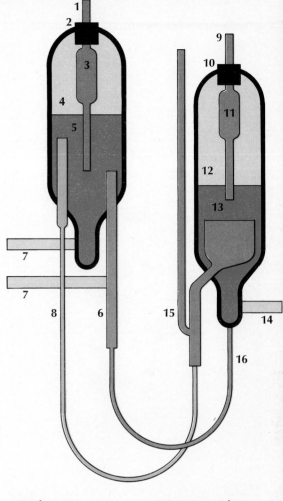

1 Stack	9 Reactor products
2 Stopper	10 Stopper
3 Flue gases	11 Flue gases
4 Overflow well	12 Overflow well
5 Catalyst level	13 Catalyst level
6 Riser	14 Steam
7 Regeneration air	15 Oil feed
8 Regenerated catalyst	16 Spent catalyst

In this catalytic cracking system, hot long-chain oil molecules mix with the catalyst in the chamber on the right. The cracking reaction leaves tars on the catalyst particles. These tars are removed and the catalyst reheated in the other chamber. The regenerated catalyst is then recycled, with fresh catalyst being added periodically.

Two hundred years ago our energy needs were satisfied almost entirely by renewable sources — wood for fuel, and wind and water power to drive our engines.

100%

75%

50%

25%

set cracks the long chain hydrocarbon molecules into fragments. A second set, usually a combination of platinum and cheaper materials mixed to form pellets, helps convert the fragments into branched chain molecules. Hydrogen is added to avoid the formation of double bonds between carbon atoms. This type of reforming can be done at relatively low temperatures, but very high pressures are needed. Equipment is needed that can withstand pressures as high as 2000 pounds per square inch.

One hundred years ago lamp oil and lubricants provided just about the total market for refinery products. With the advent of electricity, the market for lamp oil all but disappeared. Automobiles have created a market for oil refinery products of enormous size. Business operations, in turn, motivated chemists and chemical engineers to see what kinds of molecules could be made by using crude oil as the raw material.

A finite supply—an infinite demand

We are using energy faster that it is being replaced. In fact, 95 percent of the energy used today comes from deposits of coal, oil, and gas that were created millions of years ago. Thus, only 5 percent of the energy we are using comes from renewable sources and the yearly demand increases by 3 to 5 percent.

Since it is difficult to estimate accurately the total natural reserves of coal, oil, and gas, estimates of how soon these energy resources will be used up differ widely.

The amount of world coal that can be recovered by today's mining methods is estimated at 5×10^{12} tons. Since we are using coal at an annual rate of increase of 4.1 percent, the world's supply will last about 110 years. However, if the amount of world coal is actually five times greater than current estimates but the estimated rate of increasing use is correct, the world's supply will last 150 years.

Using similar logic, the world's supply of oil and gas will last 20 to 50 years. The problem of maintaining adequate energy supplies throughout the world provides real challenges as we face the future.

Using hydrogen to extend the world's supply of oil

J. O'M. Bockris, an Australian physicist, shows how people can respond to the challenge of an impending energy shortage. In *Science* (June 23, 1972) he describes how hydrogen could be used as a substitute for petroleum fuels in many present-day uses. Bockris' plan calls for nuclear reactors located on platforms floating off ocean shorelines. The energy produced by the reactors can be used to separate water by electrolysis into hydrogen and oxygen. The hydrogen can then be carried to distribution centers and from there to individual customers. Autos, trucks, trains, ships and, perhaps, even airplanes will use fuel cells to convert the energy from "burning" hydrogen to drive electric motors or power jets. These light-weight, highly efficient power systems will produce only water vapor exhaust and thus will be non-polluting.

Bockris agrees that his plan is quite a challenge to engineers and technicians. He also admits that many people fear the highly explosive character of hydrogen. These problems, however, might be solved more easily than those which are inevitable if gasoline pumps run dry or electrical power stations "brown out."

Using fuels before they become fossils

D. L. Klass describes his idea for heading off an energy crunch in *Chemtech* (March, 1974). His plan relies on the fact that 146 billion tons of potential carbon fuels are produced each year by the photosynthetic processes in green plants. The sun floods the earth each year with 28 times more energy than exists in the total world supply of coal, oil, and gas. Although green plants are not efficient in capturing and storing the sun's energy, Dr. Klass estimates the 146 billion tons of new plant tissue produced each year to be equivalent energy-wise to about a 150-year supply of substitute natural gas.

The organic material in trees, plants, grasses, or algae can be converted to substitute natural gas with about 35 percent efficiency, but there are problems. For example, large amounts of water must be removed before plant material can be converted to methane, the chief component of substitute natural gas. Another problem is the removal of inorganic plant nutrients from the soil. Also large quantities of waste or byproduct materials would have to be disposed of.

Large areas of land or water would have to be set aside to grow the plants needed to produce enough new carbon each year to substitute for fossil carbon. However, 200 years ago all energy was obtained from current rather than fossil fuel resources. In fact, wood provided well over half of our energy needs only 100 years ago. In 1876 the average person in the United States used about 5 gallons of oil per year, and natural gas wasn't much more than a puzzling curiosity.

Methanol may be the fuel of the future

G. Alex Mills and Brian Harney believe methanol stands a good chance to become an alternative for liquid fossil fuels. They explain (*Chemtech*, January 1974) that we already know how to make methanol from coal. Furthermore, methanol is enough like gasoline so that engines could be modified to use it. Because methanol can be blended with gasoline, the changeover from one fuel to another could be eased.

Methanol is non-polluting, burns in an auto engine better than 100-octane gasoline, and can be made from coal more cheaply than gasoline. A gallon of methanol, however, yields only half as much energy as a gallon of gasoline.

Currently, some 6 million tons of methanol are produced each year from methane and water, but methane (natural gas) is in short supply. Coal is the most logical substitute. Dr. Mills and Dr. Harney are confident that the proper temperatures and pressures can be worked out and new catalysts developed to allow coal to be converted to methanol and at a price and in quantities large enough to head off problems if and when our supplies of gasoline run short.

Gas from coal

Some researchers believe we can gain the time needed to shift from fossil fuels to new energy sources by learning how to make gas from coal. The Koppers-Totzek process for converting coal to gas is one effort in this direction. The process dates back to the 1900s before natural gas became popular. Pulverized coal, oxygen, and steam are brought together in a gasifier. Because they are introduced through burner heads at opposite ends of the gasifier and at temperatures near 3500°F, the coal is gasified almost instantly and completely. In larger units, three or four burner heads are used.

The gasified coal is mostly a mixture of carbon monoxide, CO, and hydrogen and contains about 75 percent of the energy originally in the coal. As the gas leaves the gasifier, it is washed to remove any coal ash that it may have picked up. The wash water is recycled as steam or is used to drive compressors or pumps.

If the coal contains sulfur, hydrogen sulfide, H_2S, is the major impurity in the gas. This well known pollutant can be removed by installing hydrogen sulfide absorbent systems.

A three-gasifier plant could produce 27.5 billion Btus of energy per day by consuming 2145 tons of raw coal.

Converting coal to gas underground

For more than 100 years people have wanted to obtain a gaseous fuel from coal while it is still underground. The advantages of underground gasification are described in the *Chemtech* issue of April 1974. R. M. Nadkarni, Charles Bliss, and William I. Watson point out that the hazards of mining coal are avoided as well as the environmental impact of spoil

100%

95%

Today only 5% of our energy comes from renewable sources. Fossil fuels provide 95% of the energy we consume.

75%

50%

25%

banks, slack piles, and acid mine drainage.

A patent on underground coal gasification was granted in 1909, but the first large-scale development of this idea took place in Russia after 1933. Large-scale exploratory work was done in the United States and several other countries after 1945. Serious technical difficulties have kept the idea from catching on.

All of the methods tried so far require the underground coal seam to be at least partially fractured or opened. This is done by explosives or by boring holes or cutting shafts into the coal seam. Air or pure oxygen is then forced into the fractured coal to convert the coal to a gaseous fuel.

All methods must be controlled carefully to ensure that the coal is not burned completely to carbon dioxide. Since the volume of the reacting "container" is always changing, so is the rate of heat escape. Ground water can introduce more problems; faults and fractures in overlying rock strata cause rock-falls or collapsing roofs.

In one somewhat typical experiment done in England, a shaft was dug down to the coal seam, and enough coal was removed to open up a room. Four parallel 14-inch holes were bored 75 feet apart and 400 feet into the coal seam. Holes were drilled from the surface to connect with each of these four holes. Air was then pumped from the room to ignite the coal along the four holes sufficiently to create gaseous fuel.

This system operated for 118 days and gasified 900 tons of coal. About 15 percent of the coal was left as supporting pillars between the boreholes. Approximately 70 percent of the energy in the coal was recovered in the gaseous fuel. After cleaning, the gas was suitable for fuel.

Cleaning up coal

Some coal contains so much sulfur that it cannot be burned without polluting the air with sulfur dioxide. Environmental restrictions have forced many electric-generating companies and other commercial users of coal to switch to fuel oil or natural gas.

The Battelle Memorial Institute of Columbus, Ohio, has devised a way to take coal apart and extract not only the sulfur but several contaminating metals that add to air pollution. Raw coal is ground and mixed with a solution containing sodium and calcium hydroxide. After heating at approximately 300°C for 30 minutes, the hydroxide ions combine with the impurities to form solids which can be filtered off. After the coal is washed and dried, it is ready to use.

Enough energy, sure, but it takes more than energy

The advance of civilization depends equally on muscles and brains, muscles to control the flow of energy through the environment and brains that will enable our muscles to control ever increasingly large amounts of energy. This brings other problems.

The more energy being transmitted, the stronger must be the lever or wheel,

The higher the temperature, the more heat resistant must be the alloy,

The greater the impact, the more elastic must be the rubber,

The greater the speed and friction, the harder must be the gears and bearings;

the "oilier" the oils and greases,

And some materials must combine many of these highly specialized characteristics.

Energy makes new materials possible

The science and art of chemistry began when people learned that energy, especially heat, can be used to take things apart and put new things together.

Long before atomic theory took shape, chemists were puzzled by the precise recipes needed for the compounds they made in their primitive laboratories. Stephen Hales in 1727 was so impressed by the precision of chemical recipes that he quoted a biblical passage which reads: "Thou hast ordained all things by measure and number and weight."

Joseph Proust described this behavior in 1799 as the law of definite proportions. John Dalton advanced atomic theory by a giant step when he suggested that all elements existed as tiny, indestructible particles. He believed that the atoms of each element have unique weights. Furthermore, when elements combined to form compounds, atoms of one element joined with atoms of other elements. Thus, if the recipe for a compound always called for precise weights of each element, it was because the recipe for the total compound simply followed the relative weights of the atoms that combined to make the compound.

A second important realization was that an equally precise amount of heat was always involved when a specified weight of a compound was put together or taken apart. This observation suggested strongly that energy, especially heat, played an important role in linking atoms as compounds.

The United States was growing as a nation at the same time Dalton's atomic theory was becoming the basis for modern chemistry. Dalton's theory explained how precise weights were involved in recipes forming compounds, but it didn't explain why atoms link to form molecules. Through much of the 1800s, chemists called this linking property affinity, but no one could relate affinity to other atomic properties. It is like calling the tendency for unsupported objects to crash to the earth gravity. Giving the phenomenon a name diverts our curiosity even though the falling remains unexplained. (Many of us would still like to know exactly what gravity is.)

The most satisfying idea that accounts for the affinity between atoms grew out of investigations of static electricity. For a long time it has been known that bits of paper, hair, cloth, in fact, tiny particles of almost anything will cling to glass or amber that has been rubbed with silk or fur. This stickiness was called electrical charge and was believed to come in two forms, positive and negative. Charged objects would attract other objects with the opposite charge.

Because this kind of stickiness was exhibited by nearly all materials and because all atoms seemed to be able to stick together to form molecules, it is easy to see why electrostatic charges became a popular way to explain molecular formation. It took more than 100 years before anyone was able to take apart Dalton's indestructible atoms and put them together in a way that would give them positive or negative electrical charges.

The atomic model that was advanced during the early 1900s leans heavily on small bundles of negative electrical energy called electrons as the outer portion of atoms. Equally small bundles of positive electrical energy form the inner nucleus of atoms. Atoms link together because the positive areas of one attract the negative areas of the other. Actually the affinity between atoms is a form of energy.

Energy can be put into atoms by forcing electrons to move farther from their positively charged nucleus. If an "excited" electron becomes caught in the higher energy level of another atom, two things happen. First the two atoms join, and secondly the new bond keeps at least some of the energy that originally boosted the electron away from its nucleus. We can retrieve the energy by taking the molecule apart and allowing the excited electron to return to its "ground state."

Nearly all of the energy that keeps us doing the things we do comes from temporarily excited electrons jumping back to their ground states. If we add the energy that is packed into substances by increasing the motion of their molecules, we can account for just about all of the energy in the world.

There is much mystery still hidden in this model of the atom where

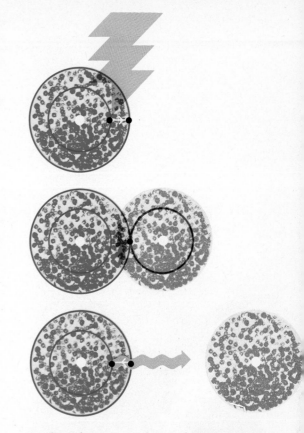

Bundles of energy can boost electrons to energy levels farther from an atom's nucleus. When these excited electrons fall into vacant spaces in other atoms' energy levels, the atoms become bonded together in energy-rich molecules. When the electrons fall back to their ground state, the molecule comes apart and the energy is released.

magic rubber bands can be stretched and snapped back to store and release energy. Energy is more easily felt or experienced than defined. We can continue to study energy and its effects. The important thing is to learn how to manage our earth's energy resources effectively.

The energy in a spoonful of sugar

A spoonful of sugar will illustrate how energy is involved in one kind of molecule. A teaspoon of sugar contains about 6×10^{21} sugar molecules. Each molecule contains 12 carbon, 22 hydrogen, and 11 oxygen atoms. A small amount of energy bonds each atom to one or two other atoms and thereby holds the sugar molecule together. Many of these bonds involve electrons that have been boosted from their normal energy levels. The source of this energy is the sun's light falling on the cane or beet that produced the sugar.

A teaspoon of sugar yields about 16 food calories, equivalent to 16,000 "chemical" calories. A total of 12,000 calories are packed into the bonds that hold the 6×10^{21} molecules together in a teaspoon of sugar. This much energy is left over when the sugar molecules are broken down in our bodies and their atoms rearranged to form 72×10^{21} carbon dioxide and 66×10^{21} water molecules. To "burn" the sugar and release this energy, 72×10^{21} oxygen molecules had to join the sugar molecules.

Similarly a gallon of gasoline contains 16×10^{24} hydrocarbon molecules. The energy released by burning a gallon of gasoline can drive an automobile about 15 miles. In other words, about 30 million calories of energy are left over when 16×10^{24} hydrocarbon molecules are taken apart with oxygen and rearranged to form 112×10^{24} carbon dioxide and 128×10^{24} water molecules.

Energy is packed into sugar and hydrocarbon molecules in very much the same way. In both cases it can be released by burning. This doesn't mean that 1000 teaspoons of sugar will drive a car 15 miles or that we can sprinkle gasoline on our cereal, but our mental model of atoms helps us see how energy can be packed into the molecules and released when needed.

Oxidation, takes molecules apart and releases the energy we need for life's activities.

82

Chemistry plays a role in making solar energy available

Solar energy is often cited as a possible solution to our energy supply problems. We assume it is perpetual and are reassured about its abundance when we read that the sun floods the earth each year with more energy than is available in the total world supply of coal, oil, and gas. In fact green plants trap enough of the sun's energy each year to equal a 150-year supply of substitute natural gas.

Before the sun's energy can do more for us than tan our skins and warm our environment, it must be trapped and transformed. This is why green plants are so vital to life. People are challenged to find ways to lend green plants a hand at trapping and transforming solar energy.

Norman C. Ford and Joseph W. Kane describe in the October 1971 issue of *Bulletin of the Atomic Scientists* a plan to harvest the sun's energy. Their proposal uses mass-produced plastic lenses built into a collecting system more than a square mile in area. The collected solar energy is focused on appropriately designed boilers which heat water to about 1500°C. At this temperature, the water molecules split into hydrogen and oxygen. To separate hydrogen from the hot steam, the boilers have windows covered with a membrane that has millions of tiny pores in each square inch of surface—holes so small that only hydrogen molecules can diffuse through. Hydrogen gas is an almost ideal fuel for various energy converting systems.

Another solar energy system planned by Aden and Marjorie Meinel uses large areas of very thin films of certain metals. When the sun's energy is reflected between these metallic films, it is absorbed, and temperatures as high as 500°C are reached. The Meinels plan to store the energy that creates these high temperatures by using it to melt substances in much the same way that heat is stored in melted ice. To match the energy output of a typical billion-watt electric power station, the Meinels' system requires a collecting surface approximately 74 miles on a side. This type of solar energy system would be located in desert areas. If abundant sources of cheap energy are available, perhaps water could be brought in and used for irrigation.

Dr. Peter E. Glaser, a mechanical engineer at the Arthur D. Little re-

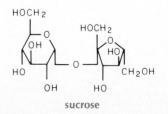

sucrose

search organization, plans to establish a space power station to collect solar energy. Two satellites would be located in space so that at least one is in sunlight at all times. A collecting disc 3.3 miles in diameter would trap enough energy to supply power for much of northeastern United States. Electronic devices would beam the collected energy toward the earth as microwaves. An antenna 1.86 miles in diameter at the space station and a receiving antenna on earth the same size would be needed. Such a space power station is a real challenge to scientists.

Molecules can store or release electrical energy

Dry cells and storage batteries are convenient packages of small amounts of energy. The flick of a switch causes energy to flow through dry cell-equipped transistor radios, pocket calculators, automobile starters, flashlights, mechanical toys, and other gadgets.

One of the first dry cells was put together in 1865 by Georges Le-Clanche, a French chemist. He put zinc and manganese dioxide, MnO_2, in a container with ammonium chloride, NH_4Cl, dissolved in enough water to make a paste. He knew this combination of chemicals would cause what we now know as a stream of electrons to leave the zinc and move to the manganese dioxide. Because of the way the cell was built, however, the electricity had to flow through a circuit outside the dry cell.

LeClanche's original dry cell was improved by using zinc to make the container and putting a carbon rod in the center. The pasty solution was packed between the zinc and the carbon rod. Other improvements replaced the ammonium chloride with potassium hydroxide, KOH. New material was found to make the container, and the zinc became a porous rod near the center of the dry cell. This solved the problems which occurred when dry cells were left too long in a camera or flashlight and the chemicals continued reacting until the zinc container was eaten away and the moist chemicals leaked from the dry cell.

Dry cells and batteries cannot provide electricity as cheaply as electric power generating systems. Power company electricity may cost no more than 5 cents per kilowatt hour, but this much electricity from an ordinary dry cell would cost nearly 50 dollars. The advantage of dry cells and batteries is that they can deliver small quantities of electrical energy wherever needed. One of the newest uses for dry cells is in wristwatches. Another increasing use is in pocket calculators and more sophisticated instruments.

Newer dry cells use solid silver iodide, AgI, to cause the electrons to flow and lead and silver chloride for the electrodes. A battery of these cells can produce a voltage of 90 to 100 volts. The rate of electron flow is quite low, but this battery has a practical life of up to 10 years.

Another new type of dry cell uses zinc and manganese dioxide like the original LeClanche dry cell, but the chemicals which react are embedded in plastic. This cell never leaks and has a long life. A 25-cell stack forms a battery no larger than a quarter of an inch in diameter and a . third of an inch long. It has a voltage of 37.5 volts and delivers current in millionths of amperes for many years.

The lead storage batteries used in automobiles are convenient sources of electricity. When these batteries are being charged, an outside source of energy changes lead sulfate, $PbSO_4$, to lead or lead dioxide, PbO_2. The electrolyte is sulfuric acid, H_2SO_4. In effect, the electrons are forced to occupy higher energy levels during charging than they had occupied in the lead sulfate. During discharge or while the battery is being used, electrons return to lower energy levels but they must travel through external circuits. Thus, they can be made to start automobile engines, light electric bulbs, and ignite spark plugs.

Rechargable batteries are always being improved. The silver–zinc storage cell, for example, uses a porous zinc plate containing a little mercury for the negative electrode. The positive electrode is a silver screen covered with silver oxide. The electrolyte in this cell is a solution of potassium hydroxide, KOH, that is saturated with zinc hydroxide, $Zn(OH)_2$. This storage cell can be charged and recharged hundreds of times. Because they are so convenient, dry cells and batteries are being used more and more often as energy sources.

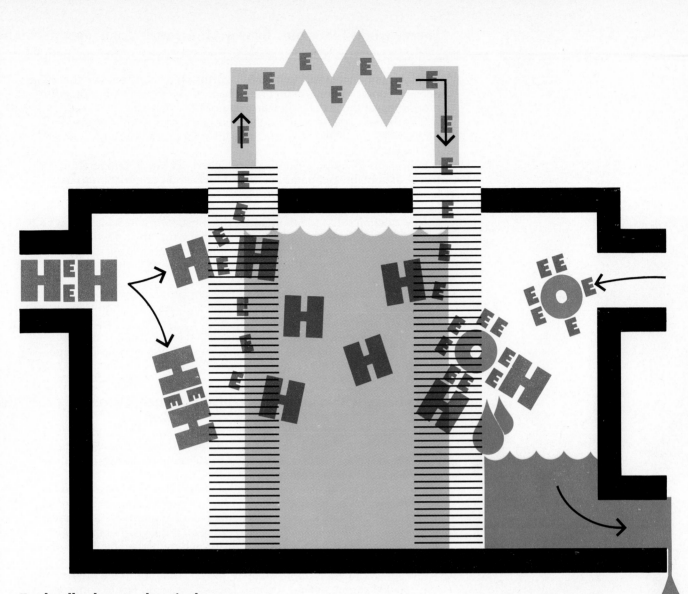

Fuel cells change chemical energy directly to electrical energy

When fuels burn, electrons move from higher energy levels in fuel molecules to the lower energy levels in carbon dioxide, water, or other end products. This happens as heat energy is liberated. Chemists are challenged to find ways to capture this energy.

Fuel cells convert the energy of moving electrons directly to usable energy. Interest in fuel cells grew rapidly in the early 1960s because spacecraft need highly efficient energy sources. A hydrogen–oxygen fuel cell, for example, not only produced energy efficiently, but the "waste" product, water, was also useful.

In this kind of fuel cell, the hydrogen is fed to the surface of an electrode where it reacts with oxygen to produce hydrogen ions and electrons. The hydrogen ions travel immediately through an electrolyte to another electrode where they absorb electrons from oxygen to produce water. If the two electrodes are connected by an outside circuit, electrons will move to balance the electron transfer that occurs inside the fuel cell. When these electrons flow they can be made to do work.

Clever engineering can manipulate electrons. In one type of fuel cell the electrodes are formed from a mixture of powdered catalyst, usually platinum, and a wetproofing agent. The mixture is embedded in a woven metal screen. The electrodes are grooved so that hydrogen and oxygen will flow across them and allow the resulting water to evaporate.

Fuel cells promise to become increasingly popular energy sources. Cheaper substitutes for platinum and improvements in their efficiency and ease of operation continue to be a challenge.

Fuel cells not only "burn" fuels without wasteful flames but they also make electrons do work when they move from the fuel to the end product. Here hydrogen molecules are stripped of their electrons by a platinum catalyst. Before the hydrogen ions combine with oxygen to form water, the electrons must travel through an external circuit where they can be made to do work.

Energy works for us only by way of materials, tools, and machines

New materials for the future call for:

Alloys with increasing springiness, corrosion resistance, lower or higher density, good or poor conduction of heat or electricity, harder or softer properties, or those that will take and hold sharper edges,

Plastics with greater transparency or opaqueness, harder or softer properties, chemically resistant or easily biodegradable, highly colored or colorless,

Lubricants that won't evaporate at higher and higher temperatures or freeze at very low temperatures, won't lock when called upon to carry heavier loads, or won't leak through gaskets and seals when spun at higher speeds,

New ceramics that stand up under the most corrosive acids, that won't chip or crack when given severe shocks, that won't shatter when subjected to even greater and more abrupt temperature changes,

New adhesives that will fasten all kinds of materials with a strength never before equalled,

Paints and varnishes to protect all kinds of surfaces and new packaging materials that will protect against damage and contamination from increasingly hazardous conditions,

Dyes and pigments whose molecules are stable regardless of how much light energy threatens to rearrange their atoms or how long the exposure.

The energy within atoms may solve our energy shortage problems

Splitting the atoms in a weight of uranium no greater than that of a 5-cent coin releases as much energy as is released by burning 15 tons of coal. There is enough heavy hydrogen in 10 gallons of seawater to produce this same amount of energy.

The story of the enormous amount of energy that can be released by nuclear reaction began with Albert Einstein's famous equation, E equals mc squared. This equation says that energy and matter are interchangeable. When matter (m) is changed to energy (E), the yield of energy is equal to the mass of matter being transformed multiplied by the speed of light (c) squared.

The speed of light is approximately 186,000 miles or in the metric system 300,000 kilometers per second. This is almost too fast to comprehend. A flash of light could go around the world seven and one-half times in one second or 40 times while you read the previous sentence. Although it takes days for high-speed spacecraft to reach the moon, a flash of light travels from the moon to the earth in less than two seconds.

Splitting or fissioning atoms is one way to release their energy. Alternatively, in atom fusion, low-mass atoms are brought together, or fused, to produce larger, more compact atoms. When the mass of the fused atom is less than the sum of the masses of the smaller atoms, the lost mass becomes energy, in keeping with $E = mc^2$.

Hydrogen atoms are the simplest atoms. Most hydrogen atoms consist of a single proton and an electron. A small fraction of the atoms in any sample of hydrogen also have one or two neutrons in each atom. These isotopes, deteurium and tritium, can be fused to form helium, and this is the fusion reaction of the hydrogen bomb.

The energy yield from atomic fusion stretches our imagination. The deuterium in one quart of seawater would produce enough energy to meet the needs of an average person in the United States for many days even though each person's daily share adds up to 240 million calories.

Extremely difficult technological problems must be solved before fusion energy can be harvested and distributed. Before deuterium atoms can be made to fuse and release energy, the fuel must be heated to temperatures above one million degrees. Once the reaction is started, temperatures far above this develop. No materials presently available can withstand these temperatures.

In "The Necessity of Fission Power," which appeared in the January 1976 issue of *Scientific American,* H. A. Bethe discussed many of the choices and decisions involving our energy resources which must be

Albert Einstein

Out of the brilliance
of Albert Einstein's mind
was created the theory that
any bit of matter (m),
moving at the velocity of light (c),
in effect, becomes energy (E).

E=MC²

FISSION

The energy accompanying atomic fission is
equivalent to the difference between the
combined mass of the neutron and the
atom being split and the mass of the combined fission
products. If 2.2 pounds of uranium-235 is fissioned, the
mass that is lost is roughly equivalent to the energy
developed in the explosion of 20,000 tons of TNT.

Neutron
1.009 amu

U-235
235.118 amu

Mo
94.936 amu

Neutron
1.009 amu

.223 amu becomes energy

La
138.950 amu

Neutron
1.009 amu

FUSION

In atomic fusion, low-mass atoms such as
the heavier isotopes of hydrogen are fused
to produce smaller numbers of higher-mass
atoms. The energy that is released is
equivalent to the difference between the
combined mass of the atoms being fused
and the mass of the fusion products.

2.01474 amu
₁H²

⁴₂He
4.00387 amu

.02561 amu
becomes energy

₁H²
2.01474 amu

faced within the next few years. Dr. Bethe received the Nobel prize in physics in 1967 for his discovery in the late 1930s of the nuclear reactions which are responsible for the energy of the sun and other stars.

If the United States must find sources of energy other than gas, oil, and coal, Dr. Bethe believes that the energy that is released by the fission of uranium is our best hope. He discussed other alternatives such as solar energy, but only nuclear fission has been developed sufficiently to meet our energy needs by the time supplies of gas, oil, and coal become critical.

Chemists play essential roles in making available the energy that is released by nuclear fission. Unlike gas, oil, and coal, uranium is not found in nature ready to use. Although uranium atoms are distributed widely in nature's rocks and soil, they are always combined with other kinds of atoms. Even the richest uranium ores contain less than 20 pounds of uranium per ton of ore, and most ore deposits contain less than 5 pounds of uranium per ton of rock.

To extract uranium from its ore, the ore is pulverized and treated with solvents which separate the uranium compounds from other minerals present. The uranium is then recovered from the leaching liquor either by addition of a solvent or by ion exchange techniques—that is, by passing the leaching liquors through columns that are packed with materials that exchange other ions for the uranium ions that are in solution.

For use as reactor fuel all impurities must be removed from the uranium through a series of reactions which precipitate practically all impurities. A more difficult step is the separation of the fissionable uranium-235 isotope from the much more abundant but nonfissionable uranium-238 isotope. In general, both kinds of uranium atoms share the same chemical properties and hence cannot be separated by precipitation of one but not the other component of a mixture.

Actually, rather than completely separating the two isotopes, uranium fuel is "enriched" in terms of the uranium-235 content. The problem is to increase the percentage of uranium-235 to more than the 0.7 percent abundance in uranium metal.

To obtain enriched uranium fuel, the uranium is converted to uranium hexafluoride, UF_6, which is solid at room temperature but becomes a gas at slightly higher temperatures. In keeping with kinetic-molecular theory, heavier gas molecules travel at higher speeds than lighter molecules at the same temperature. Because each uranium-238 atom contains three more neutrons than each uranium-235 atom, each uranium hexafluoride molecule that contains the heavier isotope moves a bit more slowly than its neighbor molecules which contain the lighter isotope.

This fact is used to enrich uranium with the lighter, fissionable isotope by the gaseous diffusion method. A typical gaseous diffusion enrichment setup consists of hundreds of chambers with each chamber partly divided by a porous barrier. The "raw" uranium hexafluoride gas is passed through a series of hundreds of these chambers. The flow is controlled so that the gas can either diffuse through the tiny holes in the barrier or go below the barrier. The lighter molecules, the ones containing the uranium-235 atoms, are more likely to pass through the barrier. On this basis, as the gas works its way through the set or "cascade" of separation chambers, the portion that flows through the barrier becomes increasingly enriched with the uranium-235 isotope. The product can be withdrawn when it is enriched sufficiently.

Solving the enrichment problem is only one of the ways chemists help make the energy of nuclear fission available. The very nature of nuclear energy and its enormous dimensions place a heavy burden on the strength, corrosion resistance, and other properties of the materials used in nuclear reactors. The problem of nuclear waste disposal is treated in Chapter 7, "Choices Through Chemistry."

With the knowledge already accumulated, it is virtually certain that all problems associated with making nuclear energy available to the general public by the fissioning of uranium can be solved. Whether or not these problems are solved before gas, oil, and coal supplies become critical is much more difficult to predict. Very complex political and economic decisions and choices as well as technological problems are involved. However, as argued by Dr. Bethe, it could well be that nuclear energy is at the very center of many of our hopes for the future.

6 Extending Our Senses: Seeing Further With Chemistry

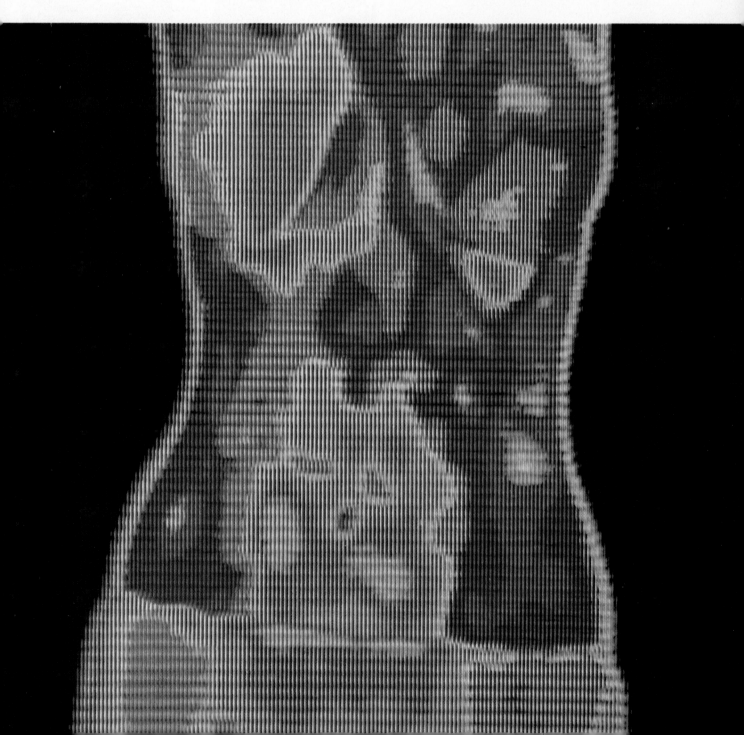

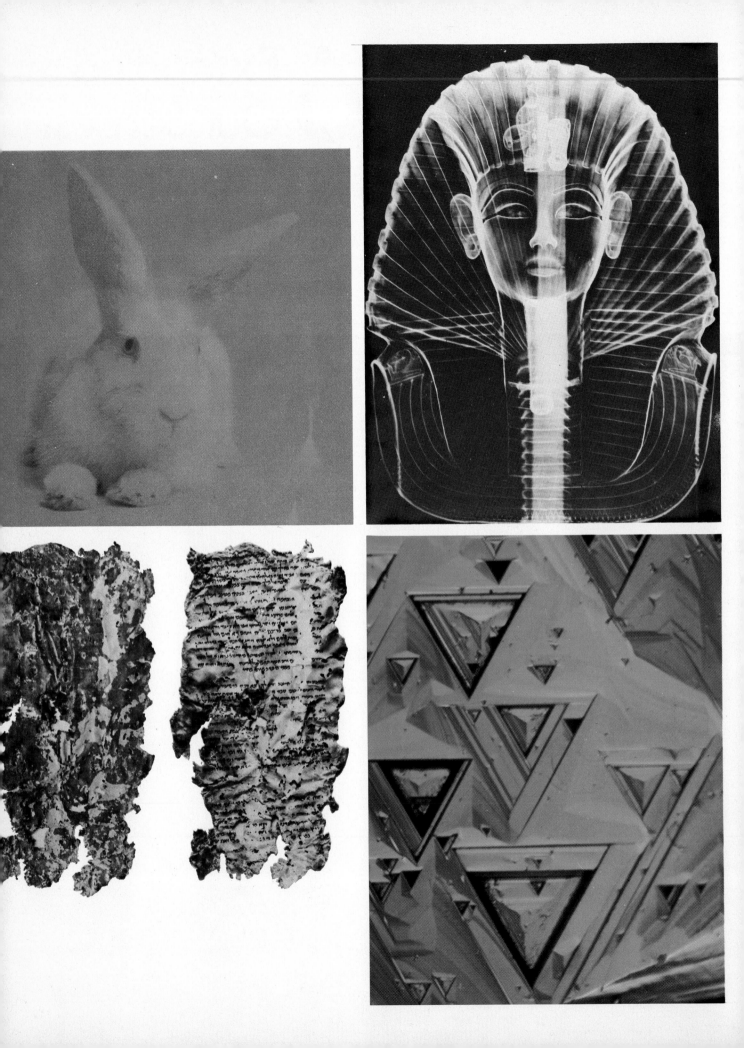

The need to communicate, to exchange ideas, to recall and record past achievements is universal. The medium can be words, paintings, or photographs. For some, photography represents the supreme achievement in the effort to share ideas and information.

Photography is a partnership between light and chemistry, a partnership with a long history. Before 1776 European chemists realized that light darkened certain substances which contained silver. Nevertheless it was nearly 70 years before Joseph Niepce, a French physicist, and Louis Daguerre, a French painter and physicist, created the first photograph.

These men covered a copper sheet with a thin layer of silver and then let iodine vapor act on the silver in the dark. This action produced a light-sensitive surface. Niepce and Daguerre were familiar with a camera used by artists to project images onto paper or canvas to aid them in preparing sketches. They used this camera to project an image onto the iodine-treated silver on the copper sheet. They hoped the silver would be darkened in proportion to the amount of light that formed the different portions of the image. Daguerre announced in 1839 that their hopes had been realized. The process was successful although Niepce died before the success was announced.

The pictures made by this process were called daguerrotypes, and although they didn't have the quality of today's photographs, they represented a giant step in the art of communication.

Within two years, William Henry Fox Talbot achieved a second giant step. He took advantage of J. B. Reade's discovery that sodium thiosulfate, $Na_2S_2O_3$, could be used to wash away the silver that hadn't been darkened when the light-sensitive surface was exposed to the light that formed the image. This solved the problem of daguerrotypes' darkening gradually after being exposed to light.

Fox Talbot's photographs were good enough to attract the first generation of camera bugs. Many of these people were scientists, engineers, and inventors in Europe and the United States. Naturally they began to design improvements.

This new invention sparked the interest of George Eastman, a name linked with photography for over 100 years. In 1874 Eastman was a bank clerk who bought a photographic outfit to take with him on a vacation. It was quite an outfit. The camera was as big as a suitcase, and it had to be supported on a heavy tripod. Also included was a light-tight tent large enough to work in while the light-sensitive chemicals were spread on glass plates before exposure and, later, while the plates were developing. Additional equipment included chemicals, glass tanks, a heavy plate holder, and a jug of water.

The glass tank which held the silver nitrate, $AgNO_3$, solution used to sensitize the plates had to be protected against breakage. Eastman wrapped it well and to give it added protection packed it with his underwear. The outcome is easy to imagine if you know that colorless silver compounds in solution darken permanently when exposed to light and spilled on hands or clothing.

This incident may or may not have been the reason Eastman decided to make picture taking easier. He became absorbed in photography and read everything available about how others were making and developing photographic plates. He was particularly impressed by the idea of using gelatin to "soak up" light-sensitive silver compounds. When the gelatin "set," the compounds remained sensitive and were in a spillproof form.

Shortly after 1876 Eastman went into the business of making dry photographic plates. Since the plates were glass and therefore breakable, he tried paper as a carrier for the emulsion—that is, the gelatin in which the light-sensitive materials were distributed. This switch solved some problems and caused others. His next idea was to coat paper first with a layer of plain, soluble gelatin. On top of this he put a second layer of insoluble gelatin that carried the light-sensitive silver. After exposure and during development, the gelatin layer that contained the image was stripped from the paper, transferred to a sheet of clear gelatin, and given a protective coating.

At this time Eastman decided to increase sales of his photographic materials by encouraging everyone to become a camera bug. To reach this mass market, he developed a new kind of camera. It was the first Kodak camera and came loaded with a roll of stripping paper film long

Chemistry teams up with light and closely related "invisible" radiations to enable us to "see" the unseeable. Films specially sensitive to infrared radiation provide thermographic images. This process can be used to thermograph a rabbit or to reveal diseased tissues in our bodies. Other emulsions can be used to reveal hidden clues that help decipher the Dead Sea Scrolls or to make the pictures taken by spacecraft more easily interpreted.

Other specially prepared emulsions team up with x-rays to produce shadow pictures deep within objects, even within the gold mask of a 14th century Egyptian Pharaoh. Magnification adds new dimensions to the things photography enables us to see. The highly magnified surface of an industrial diamond, for example, helps explain why diamonds are such effective grinding and cutting tools.

Photography needs many assists from chemistry.

Using the light reflected by the scene being photographed, the camera lens projects an image onto light sensitive film. The image is recorded by the light's effect on the silver salts in the film.

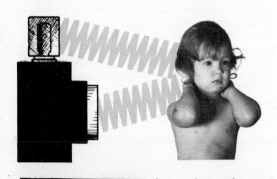

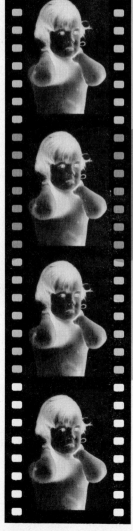

The chemistry of film development brings out the latent image in the silver salts by reducing the exposed silver particles to metallic silver.

A stop bath prevents excessive development of the exposed film.

A fixing solution removes the silver salts that have not been exposed when the scene's image was projected onto the film.

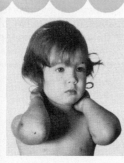

After we have washed away the chemicals that have done their work, the finished negative is dried and is ready to be used to make positive prints of the photographed scene. And this calls for more chemistry.

enough to make 100 pictures. After exposure, the camera was returned to Rochester where the exposed film was developed and printed. The camera was then reloaded with film and returned to the customer.

Since Eastman wasn't satisfied with the paper base for the light-sensitive emulsion, he hired a young chemist to find a better material. This led to a transparent plastic film made by dissolving nitrocellulose and then letting the solvent evaporate. Because it was transparent, this material could remain as the permanent support for the photographic negative and thereby avoid the cumbersome stripping process. In 1891 the film was wound on spools that could be loaded into the camera in daylight. Now cameras did not have to be sent to and from Rochester for reloading.

Improvements continued in both the film and camera design. By 1900 anyone could use this new medium of communication.

Light teams with chemistry to paint pictures in full color

Many chemistry teachers like to use what appear to be magic tricks to interest people in chemistry. There is a world of difference, however, between pulling rabbits out of hats and taking advantage of exciting changes in the properties of materials. Modern color photography, for example, is almost as unbelievable as the tricks that magicians do but it is based on straightforward chemistry.

The full story of color photography includes contributions from many people working in diverse fields. In 1611 a physicist, DeDomininis, proved that all colors of the rainbow could be reproduced by mixing the proper proportions of the three colors: red, green, and violet. In 1666, Issac Newton proved that when sunlight passed through a glass prism, the light was separated into all of the rainbow's colors. From this discovery, he argued that sunlight is a mixture of all colors of light and the color of an object depended upon the portion of the sun's light which the object reflected.

Thus, by 1676 it was known that unique colors or wavelengths of light were reflected from each different color in any scene. This led directly to the question of how our eye translates the different wavelengths of light into the pictures "painted" in our minds. In 1790 Thomas Young proposed that the human eye contains three types of light-sensitive substances. Each type is uniquely sensitive to either red, green, or violet light. He argued that the eye sees three separate "pictures" but these three pictures are blended together in the mind.

About 60 years later James C. Maxwell translated this idea into what was probably the first color photograph, a remarkable feat since black and white photography was still quite new in the 1850s. Maxwell first arranged a colorful scene. He then photographed it with ordinary black and white film, but he covered the camera lens with a sheet of transparent material that had been dyed red, in other words, a red filter. He then photographed the same scene using a green filter. He used a violet filter for a third photograph.

After the three films were developed, each was used to make a positive image of the scene on transparent material. By using three projectors simultaneously, these three transparencies were overlapped on a screen and the lenses of the three projectors were covered with corresponding red, green, or violet filters. The three separate images of the original scene were brought together to form a picture in natural color. Actually, people have had trouble reproducing Maxwell's experiment. Either he did not publish all significant details, or he was fortunate in some unknown respect. Students who have repeated his work have suggested that the spacing of the three projectors seemed to make a difference.

Inventive people immediately tried to combine Maxwell's idea with the theory that the eye contained three different light-sensitive substances and thereby solve the problem of color photography. The problem was well stated by Collen in 1865. He sought three substances, each sensitive to one of the fundamental colors of light—three substances that could be spread in thin layers on a single negative which, after ex-

posure, would form three superimposed colored images of the photographed scene.

Collen laid at chemistry's doorstep the challenge to invent the required three materials. Because of an unexpected discovery, success was prompt. In 1873 Hermann C. Vogel added a yellow dye to a batch of photographic emulsion. He hoped this would reduce the tendency for light to "spill over" and produce fuzzy effects in finished pictures. He was surprised to find that the yellow dye caused yellow and green objects to show up better than other colors after the film was developed. This chance observation was interpreted to mean that the dye absorbed yellow light and the silver was especially sensitive to yellow light. The point of this interpretation was that the light-sensitive silver emulsion was more sensitive to the color of light that a dye in the emulsion absorbed than to other colors.

Based on Vogel's chance discovery, color film was prepared by sandwiching at least three layers of emulsion. Each layer contained a blue, green, or red dye mixed with the light-sensitive silver compound. When the film was exposed, the amount of each layer exposed depended on the color of light reflected from the scene being photographed.

Because silver compounds are "color-blind," if this exposed film were developed, we would have only three superimposed black and white photos. The addition of color required some tricky chemistry. A third kind of molecule was added to the emulsion in each layer of the film sandwich. When the exposed film was developed, these molecules became dyes, and the amount of dye produced in each layer depended on the amount of silver exposed in each layer.

The dyes added permanent colors to each layer of the film, but the silver compounds, both the exposed and unexposed portions, were washed away during the developing process. Thus, the final film consisted of three superimposed layers of transparent dyes. When this transparency was projected on a screen, the three layers blended to become a full color picture of the original scene.

If the sandwich of silver compounds is coated onto white paper rather than transparent film, and other adjustments made in the total process, the final picture appears in color on paper.

Chemical laboratories we use but don't see

Modern photography enables us to expose color film, develop, and print pictures in full color in fully equipped chemical laboratories that are a part of the film. In instant color systems, the light-sensitive silver compounds and the dye molecules are a part of the film for each picture. After the film has been exposed, the sandwich is forced between two rollers which break open the pods which contained "activator" chemicals. This causes the required chemicals to be spread uniformly over the film. The kinds and numbers of molecules in the layers and activating fluids are carefully controlled. All of the chemical reactions are completed and the full color picture is available in one minute or less.

By using the chemistry and physics of photography, technologists are extending the range of our senses

Photography helps astronomers and space scientists to see farther

Special cameras installed in far-ranging spacecraft add another dimension to photography as a type of communication. Here is Jupiter's "Great Red Spot" as photographed from 660,000 miles in space by Pioneer 11.

Engineers and scientists see things that move very fast, And things that happen very, very slowly

Time-lapse photography enables us to follow the action of events that happen as slowly as the opening of a rosebud. High speed photography enables us to observe events that are as rapid as the flight of a bullet.

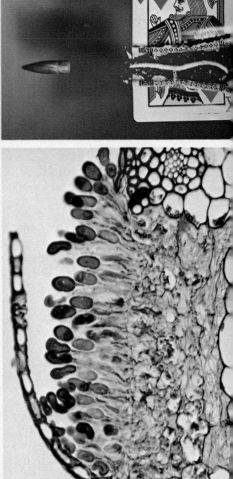

Photography even lets us see things our eyes can't see

Aided by high magnification, light can team with chemistry to reveal the detailed structure of living systems. This enlargement of the tiny parasite that causes black stem rust of wheat can help scientists control this serious disease.

Energy inside atoms paints pictures on TV screens

Two hundred years ago communication was limited to the distance smoke signals could be seen or drum beats heard. One hundred years later, distances were limited by how far wires could be strung. Today there is no theoretical limit. Seeing and talking to people on the moon proved that not even is the sky the limit. The tiny bundles of energy associated with electrons and photons travel at speeds approaching the speed of light and cover distances limited only by imagination.

In 1925 Vladimir Zworykin put together an all-electronic color television system. Forty years earlier, Paul Nipkow in Germany obtained the first patent on a complete black and white TV system. Between these two achievements is a story that may have begun when people were first entranced by the flashing light of fireflies or by the glitter of glowworms in a dark cave or by the equally ghostly luminescence of decaying wood or of strange minerals brought into a darkened room.

In 1603 Casciarolo, an Italian shoemaker, dispelled some of the mystery from naturally luminescent materials when he put together a substance that glowed in the dark after it had been exposed to light. He made this substance by heating charcoal with minerals that contained barium and called it, phosphor; this word comes from the Greek and means bearer of light.

During the 1850s H. Geissler, a German glassblower who was interested in how electricity passed through a vacuum, noticed that glass emitted a ghostly green glow when electricity jumped from an electrode in one end of a vacuum to the other. Later other German scientists proved that the glow was caused by rays that traveled through the evacuated tube. These rays could be moved by a magnet, and they caused certain minerals to glow in various colors. These rays were later shown to be equivalent to streams of electrons moving from one electrode to the other.

In 1904 and 1906 J. A. Fleming and Lee De Forest found a way to use a weak beam of electrons to control a much stronger flow of electrons. This was the first electronic tube, and it was a giant step toward the invention of TV. In 1907 Boris Rosing designed plans for building a color TV system. A glass surface would be coated with fluorescent minerals. Using this coated surface as the end of an evacuated glass tube, a magnetically controlled "pencil" of electrons would be made to move over the fluorescent "screen."

It was not until 1926, however, that J. L. Baird in England produced the first TV picture, a badly flickering, dim image on a screen only a few inches wide, but improvements came rapidly. A modern TV screen contains 200,000 precisely spaced spots of at least three different kinds of phosphorescent molecules. When hit by a stream of electrons, one kind of spot glows red, another blue, and the third green. At the base of the TV tube are three electron guns. Between these guns and the screen is a metal sieve with 200,000 precisely spaced holes. Each of the three guns can be aimed so that each can sweep the total screen but shoot its electrons only through the proper holes.

At the broadcasting studio, immediately behind the TV camera lens are four mirrors. One mirror transmits all colors of light except red; it deflects red light toward a special type of electron tube. A second mirror does the same to all blue light. The third and fourth mirrors selectively transmit green light toward a third electron tube.

Each of the three electron tubes translates the light it receives into electrical current that varies exactly as did the intensity of the light it received. In turn, this varying electrical current is transformed into radio waves that are broadcast far and wide.

When these broadcast waves arrive at a TV receiving set, the energy transformations that took place in the camera are reversed. The radio waves are changed to electrical currents and then to beams of electrons which are shot through the 200,000-hole sieve and made to sweep the proper spots of phosphor on the screen.

Chemistry's special contribution to color TV are the phosphors that produce the color. Calcium halophosphate is a widely used phosphor that is made up of calcium, fluorine, chlorine, phosphorus, oxygen, antimony, and manganese. The raw materials are weighed, blended, and fired at more than 1100°C for several hours. If the proper proportions are used and all impurities removed, the resulting calcium halophosphate is a very good phosphor.

Much of the art of making good TV screens depends on phosphors which match the natural colors of the scenes being broadcast. A slight change in the composition of a phosphor molecule can cause a decided change in the color of light it fluoresces. TV is a precise process. Physicists and electronic engineers know how to control electrons with amazing accuracy. Each burst of electrons hits its proper hole in the sieve with each sweep of the screen. Chemistry's role is to make sure the billions of molecules in each spot give back the proper color of light.

When we think back to the invention of color TV, Vladimir Zworykin was a remarkable scientist. An anecdote suggests what kind of man he was. In the 1950s, many U.S. citizens were upset because of Russia's success in the space program. Successful Sputniks somehow suggested that science education in the United States was in need of help.

In this spirit, Zworykin, who was born in Russia but emigrated to the United States in 1919 when he was 30 years old, invited several educators to his laboratory at Princeton, N.J. He explained that everyone at that time, including his boss at RCA, was involved in improving science teaching. He explained that his Russian accent kept him from making speeches but that he knew how to make electrons do tricks. This was his way of referring to his inventions of color television and the electron microscope.

More revealing was his method of improving science teaching. One of the first things he did was to visit the toy stores in town and buy each toy that claimed to help youngsters learn science and mathematics. Then he gave these toys to young children and watched them play. This was his way to look for ideas that might let him use his ability to "do tricks with electrons" to create something that would help teachers teach science more effectively.

With somewhat guarded enthusiasm he showed us his electronic "pinball" machine that was programed to teach basic computational skills. He was completely enthusiastic, however, about the prototype highly portable TV camera his "crew-cut" assistants had helped him put together. He foresaw the use of TV as a valuable teaching aid, especially on closed circuits which demanded highly portable equipment.

Electrons from deep within atoms print numbers

Two hundred years ago schoolchildren did their homework on thin slabs of slate taken from nearby rock formations. For permanent records steel pens or carefully sharpened goose quills were dipped in ink made from natural plant or mineral sources. One hundred years ago pencils became a favorite writing tool although inkwells and steel pens were still used.

Shimmering numbers which appear almost magically as read-out on electronic calculators are almost the trademark of today's arithmetic. Electronic circuits that add, subtract, multiply, divide, and do all kinds of computations are the result of many individual contributions. Chemistry's role has been to put together materials with the required properties to make the circuits and "hardware" of the electronic calculator and computer industry.

Calculators are built around semi-conductors—that is, substances that allow electrons to flow more freely in one direction than another. Calculators also combine two substances, one that has electrons that can be boosted to abnormally high energy levels, and a second substance that cannot only be "excited" but contains in its atoms vacant spaces in one or more energy levels.

We must look into the makeup of atoms to understand the shimmering light which paints the numbers in the read-out of electronic calculators. The electrons in each kind of atom are arranged in definite "stair-step" energy levels extending out from the nucleus. If an electron moves from one energy level to another, a tiny quantum or bundle of energy is either absorbed or emitted. Electrical energy can be used to boost electrons to higher energy levels. If the proper kinds of atoms are selected, the electrons will give back this energy in the form of light when they fall back to their "ground state."

One type of "light emitting diode" can be made up of tiny thin blocks of gallium arsenide phosphide. Intricate electronic circuits join the patterns of blocks needed to paint each numeral. If the calculation calls for a figure 6, for example, a stream or sequence of bursts of electrical energy is sent to the pattern of blocks needed to show a 6.

When these bursts of energy cause electrons to move through the gallium arsenide phosphide, higher energy electrons from one atom fall into lower energy vacancies of other atoms. The tiny bundles of energy that are emitted, in turn, strike other kinds of atoms and become photons. If enough photons are produced, they blend together and the light becomes visible.

Transistors can be designed to do just about everything "radio tubes" used to do, but they are much smaller, need much less power to operate, and are longer lasting. In 1958 Jack Kilby invented a complete electronic circuit containing transistors and, perhaps, diodes, resistors, capacitors, along with all of the required interconnecting electrical conductors. The heart of this device is a tiny chip of silicon roughly one-sixteenth inch square.

Although the chip is called upon to make the thousands of computations required by a calculator, the work is shared by an almost unbelievable large number of individual building blocks. There are 900 million silicon atoms on a one-sixteenth square inch of surface and the silicon chip that is used in one of Kilby's integrated circuits is many more than a single layer of silicon atoms thick.

Molecules can remember and play back what they have heard or seen

Instant replays of existing moments of football games and symphonic performances of great musical compositions stored on reels of thin tape are possible because of a curious phenomenon—magnetism. The property of naturally magnetic chunks of certain kinds of iron ore was first used to solve problems when the Chinese build direction-finding compasses during the early days of recorded history.

The story of magnetic tape as a device to record sights and sound began in the early 1800s. Hans Oersted in Denmark and Michael Faraday in England described how electrical currents are accompanied by magnetic fields and how magnetic fields create electrical currents whenever they cut through any substance that conducted electricity.

One hundred years ago Alexander Graham Bell proved that sound waves could be changed into electrical currents which retained the unique properties of the sound waves. These currents could travel great distances and then be changed back to sound waves when a moveable iron disc "clattered" against the poles of an electromagnet. The first telephone message was achieved within months of our nation's first 100th birthday.

Three years earlier it was proved that changes in the intensity of light could be transformed into electrical currents which retained the unique properties of the light when the light fell on a sheet of selenium that was a part of the electrical circuit. In 1900 Valdemar Poulson used these and related discoveries to build the first magnetic sound recording system. He used iron wire to store the sound which was changed into electrical current by using the equivalent of a telephone mouthpiece. The electrical current, in turn, was carried through an electromagnet. When the iron wire was pulled past the electromagnet, magnetic fields were induced in the wire which "copied" the characteristics of the sound received by the microphone.

During playback these same scientific principles were applied in reverse. The recorded magnetic fields in the wire were changed to electrical currents. The electrical currents, in turn, were changed to sound waves in earphones or speakers.

Sound tape recorders first went on the market in Europe in 1935 and 12 years later in the United States. The iron or steel wires of these early recorders were soon replaced by magnetic tape. The tape was made by

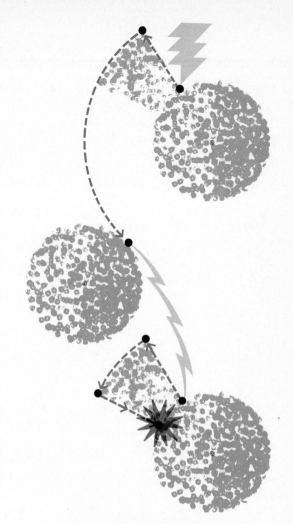

The energy from trillions of electrons "falling" from higher to lower energy levels within atoms can be transformed into the light energy that allows us to read the answers pocket calculators provide.

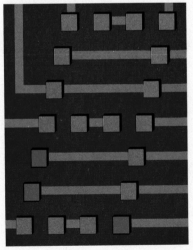

suspending tiny needle-shaped crystals of magnetic iron oxide, Fe_3O_4, in a plastic binder which was then coated onto a flexible support. To prepare the iron oxide, oxygen was added to a solution of iron sulfate, $Fe_2(SO_4)_3$, to start a reaction. Scrap iron provided additional iron and high concentration of sodium hydroxide, NaOH, helped produce the iron oxide.

Iron combines with oxygen and with water molecules to form several kinds of particles but only one kind of iron oxide particles is magnetic. When all of the required conditions are maintained carefully, pure crystals of magnetic iron oxide can be produced.

When magnetic tape is manufactured today, the magnetic iron oxide is crushed and sieved to obtain uniform size crystals. These particles are then dispersed uniformly through plastic while it is semi-fluid. If this is done properly, each iron oxide particle is separated from its neighbors. Carbon black is added to provide electrical conductivity and thereby drain off static electricity that is created when the tape rubs against surfaces.

The magnetic material is now coated onto a plastic base either by rolling, dipping, or extrusion. Webs up to 48 inches wide are coated. Eventually, these wide sheets are slit into ribbons or tapes.

Almost immediately after coating and before the plastic binding material sets, the iron oxide particles are lined up by running the semi-fluid tape past a strong magnet. When the binder solidifies, the iron oxide crystals are "frozen" in the proper orientation. Other materials are sometimes added to reduce friction and to keep the tape flexible.

Now the tape can be dried, slit, tested, cleaned, and wound on reels ready for the market. When the tape is used to record sound, a microphone and its supporting electronic devices feed electrical current into the tape that corresponds in intensity and frequency to the loudness, pitch, and quality of the sound being recorded. When used for video-tape, photoelectric cells and their supporting electronic systems feed in electrical current that corresponds in intensity and frequency to the brightness and wavelengths of light reflected from scenes being recorded.

In both cases, the magnetic particles in the tape create replicas of what they receive. During playback, these processes are reversed. The fidelity with which sounds and scenes can be captured, stored, and then recreated bears sharp witness to the talents of the men and women who devote their careers to several professions: electronic engineers, chemists and acoustical and lighting technicians.

Chemistry helps make copying and duplicating fast, accurate, clean and convenient

Not many years ago, carbon paper was about the only way to make additional copies of written material. Additional copies were made by putting a "spirit master" onto a wax-filled tray and pulling copies off one by one. Carbon paper continues to find many uses, and "ditto machines" are widely used by teachers, but new copying and duplicating processes provide reproductions at remarkably high speed, fidelity, and low cost.

Taking advantage of sticky molecules

Rather simple chemistry is used in hectograph or "ditto machine" copying systems. A dye, methyl violet for example, is mixed with wax and spread on paper to make the "master." By typing or writing on the reverse of the master, the dye is transferred to the master copy. The dye is then transferred to copy paper that is moistened lightly with a proper solvent such as methanol or 2-propanol. Copies can be made until the dye on the copy master is used up.

Taking advantage of the fact that oil and water don't mix

Direct-image offset duplicating processes also involve simple chemistry. The image to be copied is transferred to a surface that attracts water but repels oils and waxes. Oily ink is used to make the master copy. In the duplicating machine, when the master copy is rolled over an inked surface, the water-attracting surface remains clean but the ink on the master picks up additional ink which is then transferred to the copy paper. Copies can be made until the ink supply is used up or the master image wears away.

Waxy inks soaked into typewriter ribbons or spread on "carbon" paper transfer easily to form master copies. Many of the things we want to copy or duplicate, however, can't be typed or sketched easily. These shortcomings suggested a need for a way to transfer an image photographically to a master copy.

In 1830 Suckow found that gelatin that contained dichromate ions, $Cr_2O_7^{-2}$, became water repellant when exposed to light. When gelatin that contained dichromate ions was spread on paper and a photographic image projected onto the paper, the image was recorded. Treatment with additional chemicals caused the areas where light hit to attract oil whereas the unexposed areas repelled oil. This master copy or plate was used to transfer the image to copy paper in the duplicating machine.

Nearly 100 years passed before Suckow's idea was built into a successful duplicating process, but in the 1940s, machines which used the process worked so well that the copying and duplicating business grew rapidly. Such names as Lithomat, Photomat, and Plastolith became welcome additions to the vocabulary of those whose work depended on communication. Experience led to improvements both in the quality of reproduction and in the speed and convenience of the process. One outstanding development put the dichromate sensitized gelatin on a screen surface. After being exposed to a photographic image and developed, the plate served for screen printing, a reproduction process especially attractive to artists.

Thomas Edison's contribution to duplicating processes

In 1885 Thomas Edison invented the mimeograph, a duplicating process that used wax spread on a fibrous base to make the master copy. A typewriter or stylus punched the copy through the wax so that ink passed through while the copy paper stayed clean elsewhere. The process didn't catch on when it was introduced in 1887, but 40 years later the mimeograph was almost the symbol of "paperwork." By the time of World War II, the question was whether we or the enemy would be sunk by the "mimeograph fleet."

Using the fact that static electrically charged objects attract dust

The process used most widely in copying machines in 1976 began as one man's idea in 1776. In 1777, Georg Christoph Lichtenberg published his curiosity about some star-like figures which appeared on a sealing wax plate that was electrically charged. He noticed this while he was trying to discover the difference between positive and negative electricity.

In 1934 Chester F. Carlson decided that the time was ripe for a better way to copy and duplicate communications. During a literature survey, he came upon Lichtenberg's observation. Xerography was the ultimate outcome although electrophotography, a more descriptive term, was first used for Carlson's process.

This process began with a plate that holds an electrostatic charge very much like a comb picks up and holds static electricity. The plate was of such material that the charge didn't diffuse over the plate but stayed where it landed. The charge disappeared, however, when light hit the plate.

For the next step in the xerography process, an oppositely charged black or colored powder was dusted onto the surface of the plate. If the light that acted on the plate was in the form of an image to be copied, the powder stuck to the plate in the form of a "master copy." The powder that stuck was then transferred to an oppositely charged sheet of copy paper. When the powder was heated sharply, it was changed to ink that produced the finished copy.

One innovation put more of the process in the paper than on the plate. Built into the paper were heat-sensitive molecules. This idea was first developed into a thermal copying process by Carl S. Miller, but

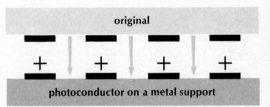

light dissipates positive electrostatic charge on photoconductor surface in non-image areas

powder adheres to the remaining charged areas

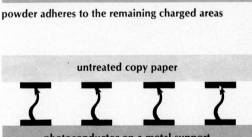

powder is transferred to copy paper and then fused

Much of the "magic" of copying machines depends upon the fundamental properties of electrostatic charges.

actually goes back to an observation published by Benjamin Franklin. He described how the sun's light and heat affect dark objects differently from light objects.

Inks

Primitive drawings on cave walls are our oldest cultural heritage. The Chinese and Egyptians used inks made from soot and animal glue as early as 2600 B.C. The Chinese were doing block printing with colored inks in 1100 A.D.

To be useful as an ink, a preparation must flow easily, dry quickly, have a satisfactory color, and keep its color as long as possible. Ink for printing must meet additional requirements dictated by the printing process and the kind of paper being used. High speed presses used to print Sunday comic sections, for example, require inks that are unsuitable for full-color material printed on coated paper such as a program for an artist's exhibit.

The raw materials for making ink are ground, mixed thoroughly, and then dissolved or suspended in an appropriate liquid. Evaporation of the solvent leaves the pigment to form the permanent message. Some of the new, highly specialized inks contain pigment in the form of small molecules. When the ink is exposed to the air during drying, these molecules are oxidized and then join together to form a continuous film of pigment.

Ball-point pen inks especially must be free flowing. In contrast, inks used for decals and designs that are transferred from plastic sheets are quite waxy. The long history of inks provides a rich heritage of knowledge that enables those who specialize in the chemistry of this form of communication to produce inks in almost every color imaginable. Furthermore, inks are available which let us communicate under every circumstance imaginable.

Paints

Thousands of test panels exposed to weather conditions during the past 50 years are responsible for the remarkable paints available today. Fifty years ago much time and effort were spent in stirring thick pigments until they were distributed uniformly through linseed oil. The paint was difficult to apply without leaving brush marks, and several days were required for the paint to dry. Cleaning up required turpentine or other highly flammable and smelly solvents.

The new paints arrived in the late 1950s with water-based latex paints. Now painting was a much simpler job, but the composition of the paint was more complex. Acrylic latex paints, for example, used several components. One popular paint consisted of methyl methacrylate, ethyl acrylate, and methacrylic acid terpolymer emulsified in water to which a defoamer and a preservative were added. Also included were a dispersing agent, titanium oxide for the pigment, and extenders to improve coverage. Finally, hydroxyethylcellulose was added to thicken the paint to make it easier to apply and ethylene glycol to keep it from freezing during storage

We must communicate

To live means to be in touch with our surroundings and with others. Communication is as essential to business as to the scientific enterprise. The entertainment world *is* communication. Nearly everything in life depends on our ability to record, measure, interpret, store, and retrieve what has been seen and heard, inferred or imagined, deduced or dreamed.

One of the greatest challenges to the science and technology of communication is the understanding of the interactions between signals transmitted by sense organs and the intricate systems of molecules and their supporting tissues in the brain. This is where education plays a strong role in communication and communication becomes a big part of education.

7
Choices Through Chemistry

Tracing beams of x-rays that have been diverted by the individual building blocks in molecules and crystals tell us much about how molecules are put together.

Chemists do what chemists do

Chemists take things apart and put things together to solve problems, to fulfill needs, and to help us understand ourselves and our world.

Industrial chemists produce materials usually in large amounts and under circumstances which yield a profit. Theoretical chemists are also interested in how the building blocks of materials come together and why they move apart. Cutting across these two branches of chemistry are the men and women who teach chemistry and those who use their knowledge of chemistry in governmental agencies. No matter where chemists work, in industry, in research or government laboratories, or on school and university campuses, they are united by the common interest of taking materials apart and putting them together.

Chemical science has helped make life better, but with some consequences and costs that we are just beginning to learn. This learning process is an exciting part of chemistry today. We all have many more, and sometimes more difficult, choices to make than people have ever had before. These choices are both personal and public. Chemists themselves do not have the exclusive right, the capacity, nor the desire to make many of these choices for us. But an appreciation of chemistry is essential to making them. For if chemistry has made some of the most agonizing choices necessary, it has also made some of the most promising choices possible.

Understanding the achievements of chemistry, the problems of chemistry, the opportunities of chemistry depends upon some understanding of the process of chemical science. Chemists and other scientists cannot always know the implications of what they learn and do. The opportunities for benefiting society are infinite. Like other scientists, chemists produce the raw material of choice. But the choices are up to all of us, individually and as communities.

The DDT story

In the early 1940's, people welcomed a highly successful new insecticide. It was nicknamed DDT, an obvious convenience because its full name was 2,2-bis (p-chlorophenyl)-1,1,1-trichloroethane. A shorter name, dichlorodiphenyltrichloroethane, explains the DDT abbreviation.

This pesticide was dramatically effective during World War II. Serious outbreaks of typhus and other insect transmitted diseases were controlled almost miraculously in countries where public health facilities had been destroyed. As soon as wartime priorities permitted, DDT was called upon to control insect pests in the United States. One of the highest priority targets was the hordes of mosquitos produced in every stagnant pond or land-locked marsh.

The swamps and marshes of Long Island were ideal breeding places for mosquitos. Since Long Island is densely inhabited, public health officials were quick to use DDT to control mosquitos. From the point of view of

A very small quantity of D.D.T. kills insect pests. The poison accumulates, however, in birds such as robins, which eat hundreds of treated insects.

This disrupts the chemistry of egg shell formation and the damaged eggs don't hatch.

Achievement

Effective chemical protection against the insects and other pests that compete with us for food and cause disease.

Choice

Hard personal decisions about whether or not to use pesticides; hard public decisions about the control of pesticides that may be hazardous.

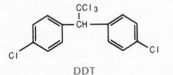

DDT

dieldrin

thousands of residents in the area, the environment was improved greatly. Nevertheless, some people worried that yearly applications of DDT might accumulate enough to be a health hazard.

George M. Woodwell, Charles F. Wurster, Jr., and Peter Isaacson decided to do more than simply worry. They took samples of plants and animals and of the muck and subsoil in a marsh at the eastern end of Great South Bay. Residues of DDT ranging from 3 to 32 pounds per acre were found in the muck and subsoil. Water samples showed DDT concentrations averaging 0.00005 part per million. This is usually abbreviated ppm and can be read as weight of DDT per million of the same weight unit of the substance or organism in which the DDT was found.

One fact especially interested Woodwell and his team. The farther up an animal was in the food chain, the greater was the absorption of DDT in its body tissues. For example, the tiny plankton organisms carried concentrations of DDT a thousand times greater than that of the water in which they lived. Birds, near the top of the food chain, carried residues a million times greater than the water.

The Long Island marshes were not the only place where DDT and similar pesticides were being used. By 1969 the nation's farmers, orchardists, herdsmen, foresters, resort owners, and public health officials were using 800 million pounds of new pesticides each year and with spectacular results. Never before were fruits and vegetables sent to market so free of worm holes and blemishes. Never before had people been able to picnic on open patios or linger over outdoor campfires without being eaten by mosquitos. Never before had resort owners been able to advertise their beaches and nature trails as free from sand fleas, jiggers, no-see-ums, and other threats to the health and comfort of their guests.

Weed and insect pests were controlled by loading herbicides and insecticides into sprayers or duster. A "crop dusters" could cover thousands of acres of fields, forests, swamps, or countryside. These new pesticides were selectively poisonous, that is, they killed only certain kinds of weeds or insects. Before selective pesticides were put together by chemists, farmers and orchardists used pesticides that poisoned all kinds of plants or animals, pests and pets or crop plants alike. When crops were sprayed with lead arsenate or Paris Green or when grain bins were fumigated with cyanide gas, everyone worried that people, especially children, would be inadvertently poisoned.

It is easy to see why selectively poisonous pesticides found ready

markets. By 1969 in the United States alone some 900 active substances
were being used in more than 60,000 brands of pesticides. At least two
pounds of one or another pesticide were used each year for each U.S.
resident. It is equally easy to see why ecologists became concerned.
These large quantities of new kinds of molecules threatened to upset the
delicate relationships among all plants and animals and their environment
including man.

In 1955 a fish hatchery on Lake George found that 100 percent of
nearly 350,000 lake trout eggs failed to hatch. At that time, 10,000 pounds
of DDT were used each year to control gypsy moths and biting flies in the
vacation areas around Lake George. In a Florida marsh where dieldrin, a
pesticide with effects similar to DDT, was used to kill sand flies, more than
a million fish died. In a city where DDT was used to protect elm trees
from the insect-transmitted Dutch elm disease, the robin population
dropped nearly 98 percent in four years. Several kinds of hawks and
eagles suffered drastic population decreases with some nesting areas
becoming vacant although they had been occupied for years. In apple
orchards where DDT was used to control coddling moths and other insect
pests, red spider mites flourished. Apparently, the DDT killed ladybugs,
but the spiders which ladybugs feed on differed sufficiently from insects
to be immune to DDT.

Situations such as these pointed increasingly toward an environmental
pollution problem. The problem was peculiarly complex and sticky. It
involved not only the complex interactions among species in the total
environment but the equally complex interactions among people with
different interests and concerns. During the 1960s much time and energy
were spent thinking through the advantages and disadvantages associated
with using the new pesticides to control weed and insect pests.

Two publications which appeared in 1969 tell the full story of the
pollution-by-pesticides controversy. One was the report of 14 people
appointed by the Secretary of Health, Education, and Welfare. Their
assignment was to gather "all available evidence on both the benefits and
risks of using pesticides, evaluating it thoroughly, and reporting their
findings and recommendations." Their report included in its title, "Pesticides and Their Relationship to Environmental Health."

"Cleaning Our Environment: The Chemical Basis for Action," was the
second publication. It reported the thinking of more than 50 people
brought together by the American Chemical Society. The hundreds of
pages in these two publications tell a story that illustrates well how chemistry and chemists become involved in the needs and concerns of all of
the people in a society. The full story is much too long to tell here, but it
teaches a valuable lesson. Supposedly, human intelligence can let us cope
with the fantastic complexity of our environment. We can investigate the
action and interactions of each factor. These interactions can be described
clearly enough to let us control, manage, or adapt their effects so we
can survive.

Research and development in pesticides is exceedingly complex. New
approaches to pest control delve into the molecules involved in the reproduction and survival of individual species of weeds or insects. A
particularly challenging part of new research seeks to identify those
molecules which control insect metamorphosis, that is, the changes which
enable larvae or "juvenile" forms of insects to become adult, reproducing
forms. Closely related is the search for equally elusive molecules which
enable one sex to attract the opposite sex at critical periods in the life
histories of insects. Because juvenile hormones and sex attractants may
impact on the total environment, their effects must be anticipated.

To have or not to have chewing gum

The choices and decisions regarding the production and distribution of materials often involve a sensitive question—that is, how closely should these choices and decisions be controlled by governmental agencies. A chance remark by a Russian tour guide illustrates the wide differences in attitudes toward governmental control. Russian children were eager for the chewing gum offered by a group of tourists. Somewhat disapprovingly, the tour guide remarked, "We don't have chewing gum in Russia, but our government right now is deciding whether or not to have chewing gum factories."

In America, people have always chewed something. The Indians chewed spruce gum, a juice or sap which oozed from spruce trees where the bark had been injured. In Central America the ancient Mayans tapped sapodilla trees and collected the milky latex sap which, when boiled, became a smooth, chewy gum. The people of Central America today chew the same latex.

Chewing gum was on its way to becoming a multimillion dollar United States industry in 1869. Thomas Adams tried unsuccessfully to make rubber from the sap of the sapodilla tree. He decided that the coagulated latex, or chicle, would be a better chewing gum than the paraffin chewing wax available at the time. It was smooth, chewy, and did not stick to the teeth.

Chicle, the basic ingredient of chewing gum, was gathered by tapping sapodilla trees. During the proper season, workers climbed the trees and made criss-cross cuts around the trunks from base to top. The latex sap flowed down the cuts and was collected in a vessel. A 70-year-old tree yielded about 2 or 3 pounds at one tapping and could be tapped every 7 or 8 years.

Just as synthetic rubber has replaced much natural rubber, synthetic waxes and polymers have replaced much of the natural chicle from sapodilla trees. For both synthetic and natural gum bases, the gum is heated until it becomes "runny," and then sweeteners and flavoring extracts are added. After thorough kneading, the gum is sent through rollers, cut into strips, sprinkled with powered sugar, dried, and packaged for marketing.

People chew gum mainly because it is fun. It is difficult, however, to imagine our governmental agencies controlling its production and distribution. How willing we are to have governmental agencies control private enterprise depends on our national traditions as well as on the properties, characteristics, and uses of the end product.

The problem of radioactive wastes

Our responsibility for waste disposal confronts us with increasingly troublesome choices and decisions. For example, one of the stickiest problems connected with using the fission of uranium to provide energy is getting rid of the radioactive wastes. Not only are the fission products radioactive, but water and other materials used around the reactor also become radioactive and must be disposed of properly.

Actually only a few people are directly responsible for the decisions regarding disposal of large quantities of radioactive wastes, but there is a lesson to be learned from radioactive waste disposal that applies to many kinds of environmental pollution. J. J. Davis and R.F. Foster traced the wastes that were flushed or escaped from the Hanford Laboratories and that appeared in the Columbia River near Richland, Wash. Circumstances enabled these men to spell out how these radioactive wastes affected the total Columbia River environment.

The Hanford story began with our increasing demands for adequate energy resources. The immediate goal was to convert the more abundant but non-fissionable isotope of uranium, uranium-238, to fissionable plutonium-239. Metal rods containing uranium-238 were put in a reactor that was fueled with uranium-235, the fissionable uranium isotope. When the uranium-235 atoms split, they ejected neutrons which lodged in the nuclei of the uranium-238 atoms. Each time a neutron lodged in the nucleus of a uranium-238 atom, two neutrons lost beta particles and

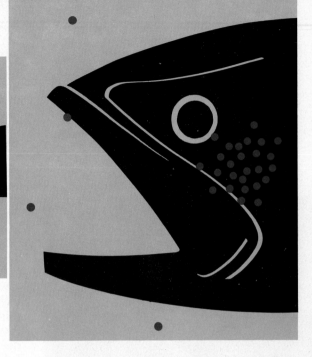

became protons. This changed the uranium-238 to plutonium-239 which could now be used as fuel for another reactor.

This breeder reactor used a great deal of water for cooling and cleaning. After its use, the wastewater which was now radioactive had to be disposed of, and this posed a problem. Teams of concerned people who represented the social, political, biological, and physical sciences established guidelines in the form of "maximum permissible concentrations." This was a "calculated risk" decision, but it permitted the vital work at the Hanford Laboratories to continue. Soon, however, Davis and Foster discovered that fish downriver from the Laboratories had absorbed 100 times more radioactive material than fish upriver. Furthermore, the concentrations in the fish exceeded the permissible concentrations for the river.

Davis and Foster studied the problem by looking at the food chain process. Tiny organisms strained the radioactive materials and concentrated them in their cells and tissues. Larger organisms which fed on them, in turn, retained more of the material than they excreted. The pattern continued and hence the largest animals in the river, the fish, ended up with large concentrations of radioactive materials.

A preliminary experiment confirmed this hypothesis. Fish were taken from an uncontaminated part of the river and divided into two aquariums. Both aquariums contained the supposedly contaminated Columbia River water. Both sets of fish received the same food, but in only one aquarium had the food organisms been collected where they had had a chance to concentrate the radioactive wastes. The fish which ate these organisms became 100 times more radioactive than the control fish but only insofar as certain kinds of radioactive elements were concerned. Some elements were no more concentrated in the tissues of the fish than they were in the river water.

Davis and Foster knew the kind and amount of each radioactive waste material in the water. Some of the radioactive isotopes in the water were phosphorus-32, sodium-24, arsenic-76, chromium-51, and copper-64. Some of these elements are important in plant and animal life processes. Others are not. Arsenic, for example, although responsible for much of the radioactivity of the wastewater, is found only rarely in normal metabolic activities and, hence, was barely detectable in the tissues of the fish.

Investigations such as these show how wastes can be traced through the environment after they have been discharged. If enough is known about the substances that make up waste and how each substance affects the environment, measures can be taken to head off possible damage.

One of the most valuable contributions of the tracer isotopes to biological research has been to reveal how large numbers of smaller living systems provide the materials that are needed to support smaller numbers of larger systems.

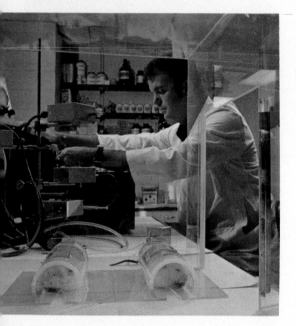

Knowledge gained from using sometimes highly sophisticated equipment and techniques on laboratory animals helps advance understanding and, hence, speeds up solutions to human problems.

Achievement

Vastly improved techniques for diagnosing, preventing, and treating diseases, controlling birth rates and correcting infertility.

Choice

With the capacity to preserve and extend human life, many face hard personal decisions about how long to sustain life and under what circumstances; and hard decisions about how to handle population increase.

When typhoid fever came to town

Amidst the death and suffering of the Viet Nam War, the manufacturer of certain kinds of chemicals used as the headline for an advertisement: IN VIET NAM JUNGLE WATER IS MORE DANGEROUS THAN THE VIET CONG. The message was that contaminated water had been killing the Vietnamese people for more than 2000 years. Typhus, typhoid, amoebic dysentery, infectious hepatitis, together with malnutrition limited their life span to 35 years. Furthermore, half of the Vietnamese children died before they were 5 years old.

One hundred years ago the diseases spread by contaminated water were almost as deadly in the United States. The incidence of typhoid and paratyphoid, for example, was close to 100 per 100,000 people, but by 1976 it was well under five cases per year for each 100,000.

Victims of typhoid give off millions of typhoid-producing bacteria. If these bacteria are spread through water or food supplies, an epidemic may develop. House flies as well as uncooked fruits and vegetables carry the bacteria, but the greatest threat lies in water supplies polluted through improper disposal of solid or liquid wastes.

The need for constant surveillance is dramatized by what happened in 1959 at Keene, N.H. This city had long been proud of its environment, including its water resources. It was one of the first cities in New Hampshire to establish a public water supply system. For nearly 100 years, except for a mild epidemic of "upset stomachs" in 1923, there was no question that the system supplied safe water to the city.

Until 1886 the city obtained its water from Goose Pond. To supply all the needed water, in 1886 the city arranged to obtain water also from Woodward Pond. In general both of these small lakes collected water from largely wooded watersheds, and the shorelines were controlled by the city. For years these two lakes provided millions of gallons of water each day for the people of Keene. Because the water was of such high quality by the time it reached the water treatment plant, only minimum amounts of protective chemicals were added. Some people wanted no chemicals to be used for fear of spoiling its natural taste and odor.

The engineers who operated the city water system relied on the U.S. Public Health Service "Manual of Recommended Water-Sanitation Practice." Regulations required at least monthly checks on the number of coliform bacteria in the water. These bacteria are natural inhabitants of human and other higher animal digestive systems, and their population is an indirect measure of water pollution.

Goose and Woodward Ponds had a long history of coliform counts well below allowable limits. High counts were recorded, however, five or six times during the late summer and fall of 1959. At this time, state and city officials recommended that chlorine be added to the water regularly, but these recommendations were not carried out.

On November 6, 1959 a six-year-old boy was admitted to a local hospital suffering from severe and recurring nosebleeds, high fever, nausea, and vomiting. Within a week, four more people were hospitalized with similar symptoms. By November 22, 14 cases recognized as typhoid were reported.

In detective story fashion, a team of local, state, and national public health officials tried to track down the source of infection. They knew that the typhoid organism could grow in many kinds of food as well as water. They also knew that every typhoid patient gives off milions of the typhoid bacteria in stools and urine. Furthermore, one in 20 typhoid victims turns out to be a carrier and continues to excrete typhoid bacteria for the rest of his life unless he receives proper medical attention.

On November 30 the investigating team visited a logging camp where three men had lived for about a year. It was located about 300 yards from Woodward Pond. Situated nearby was a barn housing two horses. At the rear of the barn was an accumulation of horse and human excreta. All drainage from the barn flowed toward a small creek that appeared to flow away from Woodward Pond. However, this creek actually emptied into a brook that connected with the city water supply system.

Dozens of clues pointed in the same direction. The three lumbermen had their blood and stools examined for typhoid organisms. When the stool cultures of one were found to contain *Salmonella typhi Type E* organisms, the investigators knew they had their answer. These organisms matched those which had claimed the 14 typhoid victims.

The lumbermen's camp was moved, the barn burned, the manure pile hauled away, and the ground was treated with chloride of lime. Last but not least all water delivered to the city was chlorinated adequately.

Chlorine has been used to make water safe for drinking for many years. It was used in England as early as 1800, especially during typhoid epidemics. Chlorinated lime, $Ca(ClO)_2 \cdot CaCl_2 \cdot xCa(OH)_2 \cdot xH_2O$, was used as a water-treating agent in Chicago in 1908, and a continuous treatment system was in operation at Jersey City in 1909.

Water is purified because it reacts with chlorine to produce hypochlorous acid, $HClO$. This acid diffuses through the cell walls of bacteria and, supposedly, destroys certain enzymes that are essential to bacterial growth. Calcium hypochlorite, $Ca(ClO)_2$, is often used rather than pure chlorine. Chlorine is highly volatile, corrosive, poisonous, and calls for special handling equipment. When calcium hypochlorite dissolves in water, hydrogen and hypochlorous ions are released. If there is too much hydrogen ion in the water from some other source, the formation of hypochlorous ions is blocked.

Those who have worked around swimming pools know that the hydrogen ion concentration in the water must be controlled carefully when chlorine is used. Actually, the amount of calcium hypochlorite or pure chlorine that is needed to treat large or small quantities of water can be determined quite accurately.

The decisions of chemists sometimes create choices for us

Manufacturing chemists are alert to new products for existing markets as well as to new markets for existing products. In the early 1960s the executives of one chemical company looked into the development of a beverage container that would be as transparent as glass but as light and unbreakable as metal. Since 43 billion soft drinks had been consumed in 1963 and since the market was increasing about 8 percent each year, they decided the container should be nonbreakable and nonreusable. Their experience with other products pointed toward a lightweight, transparent, shatter resistant plastic bottle.

It is interesting to imagine all of the questions and points of view that were considered before they decided to invest a considerable sum of money in developing a new plastic beverage container. Did such questions come up as to whether or not people should complain about carrying heavy, breakable bottles to and from beverage stores or whether they should complain about cleaning up the mess when they drop a six-pack of glass bottles? Or was the role of chemists to admit that people were complaining and to put together something that would give them less reason to?

F. D. Wharton, Jr., and J. Kenneth Craver tell the story of Monsanto Co.'s development of a plastic bottle in *Chemtech* (September 1975). These men report no discussion of whether or not the public should be given the chance to choose plastic beverage containers rather than glass or metal. The kinds of questions that were raised dealt with such things as: Can we be sure that the bottle we are setting out to develop will meet all of the technical performance requirements of beverage containers? Can we make the bottle and sell it at a price that will allow people to choose this type of container without having to pay more than they pay for other containers? How difficult will it be to recycle the materials in the new container, or should we plan to have them disposed of by burning or landfill? Can we be sure that the new bottle will not threaten the health and safety of those who must handle it and consume its contents?

Monsanto spent more than 10 years investigating these and other questions. Data were gathered from many sources. In fact, this episode illustrates the steps a company goes through when they decide to market a new product.

Consider first the performance requirements for the new bottle. Although most soft drinks are consumed within 60 days after bottling, the container must preserve the flavor longer than this in case supplies re-

$$\left[H_2C = CHCl \right]_x$$

polyvinyl chloride

$$\left[H_2C = CH_2 \right]_x$$

polyethylene

$$H_2C = CHCN$$

acrylonitrile

$$\left[H_2C = CHPh \right]_x$$

polystyrene

109

Operating from a mobile laboratory U.S. Environmental Protection Agency field engineers set up automatic water samplers to evaluate the water quality of a canal in Rochester, N.Y.

Achievement

The chemical modification of old materials, and the creation of new ones has changed the way we live.

Choice

Hard personal choices about the social cost of disposable products; hard public choices about tradeoffs in controlling the pollution of air, water, and landscape.

main on the shelf more than 60 days. It was agreed that the new container must preserve flavor for at least 180 days. This meant that the container would have to be of a material that would prevent flavoring components from being adsorbed on or from the container wall, that nothing in the beverage would interact with the container material, and that gases or vapors would not pass through the container wall in either direction. The new container had to retain its shape and show no more than 5 percent change in shape after a week at 100°F or one day at 120°F.

Because many kinds of soft drinks are bottled under pressure, the bursting strength of the new bottle was given considerable attention. It was agreed that the new bottle had to withstand pressures 10 times greater than atmospheric at usual room temperatures. Furthermore, the bottle had to be strong enough to not "explode" when it was under four times atmospheric pressure and warmed to 142°F. Other criteria required the bottle to withstand capping, stacking, and all of the usual stresses of filling, handling, and consuming.

Problems arose when existing materials were studied. To meet the flavor-retaining requirements, for example, a polyvinyl chloride bottle would need walls more than a quarter of an inch thick; a polyethylene bottle nearly four inches thick; and a polystyrene bottle more than seven inches thick. Other kinds of plastics were unsatisfactory because they cost too much, were structurally weak, difficult to form into bottles, or failed to meet other requirements.

Eventually, acrylonitrile copolymers were found to have properties which seemed to meet the required specifications. By 1969 Monsanto was ready to test market a bottle made from methacrylonitrile-styrene. Unfortunately, this bottle failed to be as break-resistant as needed. Few of the bottles survived a 3-foot drop when filled and pressurized. By knowing how the individual monomer molecules linked together to form the polymer, Monsanto's chemists changed the conditions under which the polymer was fabricated. These changes led to the production of bottles in which the acrylonitrile-styrene molecules were dimensionally oriented to produce very tough, break-resistant plastics.

The environmental impact of the new bottle posed problems of a different nature. What would be the effect, for example, of a beverage container that floats rather than sinks in water? Would discarded plastic bottles make more or less of a mess than paper cups or cans or bottles? Would it be better to recycle the plastic in the new bottles or would people prefer to recycle the actual bottles? Would it be better to plan to dispose of the new bottles by burning or landfill?

Obviously, problems such as these couldn't be solved by precisely calibrated balances, burettes, and the instruments chemists ordinarily use, but they had to be dealt with before the new plastic bottles could go on the market. On this basis, Monsanto used a strategy called "cross-impact analysis." The project director, the research director, development manager, production engineer, financial analyst, and company executives were organized into a team.

These people set out to estimate the probable effects of up to 30 "events" that might arise from the reactions of consumers, competitors, legislators, and other groups who are involved in "socially related environmental and safety criteria." As soon as these reactions had been predicted, the next step was to set the goals and establish the action that the company would undertake to keep environmental hazards and safety precautions to a minimum.

Again Monsanto chemists were involved in the plastic bottle decision-making situation because it was assumed that the public should be allowed to choose whether or not to consume soft drinks. However, if soft drinks were to be consumed and distributed in small, no-deposit containers, then it was the proper business of chemistry to make available the best possible container—the best possible, in this case, involved not only scientific and technological criteria but also economic, social, and political.

Here are some examples of the many pieces of information fed to the Monsanto team while carrying out their "cross-impact analysis" of the new plastic bottle problem. Assume that 10 million ounces of a beverage are to be distributed in 10-ounce one-way containers. If aluminum cans were used, 20,500 pounds of pollutants would be added to the air, 8,500 pounds of pollutants to the water, and 114,000 pounds to solid industrial wastes. If steel cans were used, all of these figures were cut approximately in half.

If glass bottles were used, 17,500 pounds of air pollutants would be produced, 101,500 pounds of water pollutants, and 7,000 pounds of industrial solid wastes. Comparable figures for plastic bottles were 14,500 pounds, 3,500 pounds, and 7,600 pounds.

If containers were recycled, 700 Btu's were needed to recycle the material in a plastic bottle, 1100 Btu's to recycle the aluminum in an aluminum can, 2500 Btu's for a steel can, and 2900 Btu's for a nonreturnable glass bottle.

If plastic bottles were burned in a properly operated incinerator, they would be almost completely consumed. Not until the plastic bottles became more than 8 percent of the total material being burned was there a minor change in the makeup of the gaseous emissions significantly affected.

When test samples of the plastic as well as the bottles were buried in a compost pile, immersed in a lake, buried in an open field, and exposed in the open for 116 days, neither the plastic nor the bottles showed any signs of biological degradation. These facts prompted Monsanto to consider the possibility of taking the plastic apart again and trying to put its molecules together in a way that would favor biological degradation. They decided against this action, however, because they believed it would be better to keep the plastic in a condition that made it easy to keep track of until recycled or disposed of.

When filled and pressurized plastic bottles were dropped from a height that assured they would break, practically none of the fragments penetrated a waxed paper "cage" 24 inches from the point of impact. As much as 12 percent of the fragments from a glass bottle penetrated the waxed paper. Furthermore, the glass fragments cut with one-seventh the pressure required from a plastic bottle.

Rats fed hamburger cooked over fires fueled with plastic bottle material for four weeks showed no ill effects. While it is unlikely people will eat the plastic bottle material, rats fed with 10 percent of their feed composed of the ground plastic showed no ill effects. Strips of the plastic were implanted in the muscle of rabbits along with samples of the plastic used for surgical implants. No differences between the two materials could be observed when the wounds healed.

Mr. Wharton and Mr. Craver close their story about Monsanto's new plastic bottle by pointing out that "efforts at technological assessment are continuing." This, too, is characteristic of situations in which chemistry makes available new solutions to problems. In fact, no matter how much effort or money is spent trying to predict how a new product will measure up to the demands of the market, if the product is to be successful, those who manufacture and distribute it must watch consumer reaction closely. Sometimes this is the only possible source of vital information.

Chemistry helps to broaden our choices involving the food we eat

Nearly 2 million people's jobs involve processing or preserving the 8000 different kinds of food that await consumers on supermarket shelves in the United States. The food industry creates about 5000 new items each year. Only one in 10 of these new items, however, survives for as long as a year.

The food industry knows that their most successful items taste good, are easy to prepare and serve, promote health, and have flavors and aromas that compare favorably with farm-fresh fruits, vegetables, or meats. Successful food products demand men and women who are knowledgeable about proteins, carbohydrates, fats, vitamins, minerals, and the even more complex substances which give foods their flavor and textures. They also must know the properties and characteristics of the "raw" foods and how to blend and season the ingredients in each recipe.

Fifty years ago our choices of foods were broadened primarily because the food industry found better methods of preservation. More recently,

Aluminum   Steel   Glass   Plastic

■ Air pollution

■ Water pollution

■ Solid wastes

How each type of container contributes to environmental pollution is one of the questions that must be considered when deciding what kind of beverage container to use.

the emphasis is on convenience. Ready to eat foods, instant soups and beverages, and mixes for puddings and baked goods that require nothing more troublesome than the addition of an egg, milk or water have become very popular.

Today food is involved in questions which call for choices and decisions. Some of these questions involve our responsibilities only to ourselves; others involve our families, neighbors, countrymen, and the whole world. Questions may deal with the kinds of food we select or the amount we can afford for our next meal or they may deal with decisions which could affect whole populations for years to come.

To gain perspective, a student reviewed the history of food processing and preservation, and to show how young people view this topic, his paper is presented here with very little editing.

Let's begin our excursion back in time in the year 1776. The colonists were faced with the problem of how to keep their food from one season to another. What these people had to work with were things passed down through the ages. They knew that salt when sprinkled on meats and fish would keep the meat from spoiling; that drying or smoking, or using spices would also enable them to keep their meats longer.

These ideas about preservation were about all the people had to work with at the time. Chemistry was too hung up with atoms and molecules to have time to fiddle with food preservation. Science on the whole was explaining food spoilage by using the theory of spontaneous generation, that is, molds, yeasts, and putrid growths on spoiled food arose spontaneously from the food.

If we move ahead to 1876, we see how this generation of people solved the problem of preserving food. This was a great century for chemistry. Chemists now had time to settle down and look at the problem. By now, Francois Appert, a Frenchman, had introduced canning as a way to preserve food. The need arose when France needed a better way to transport food to her army. Appert came up with the idea of heating foods and then putting them in tightly sealed containers. Not knowing exactly why it worked, people put their trust and their food in glass jars and bottles with tightly sealed wire and cork stoppers.

But people asked why food spoiled or kept. In 1860, Louis Pasteur answered them. From his research and investigations, he demonstrated that ferments, molds, and other forms of spoilage were caused by minute organisms found in air, water, and food. These organisms needed definite conditions of heat, air, or moisture to live. If conditions permitted them to live, they spoiled foods. These discoveries opened the way for new advances in food preservation.

Now people could preserve food by canning along with the other methods of preserving food from the past and they understood that in each of these methods they were preventing the growth of organisms which, in turn, kept the food from spoiling.

When we jump ahead 100 years to food preservation today, a new problem has arisen, a bigger population. New and better ways are needed to keep food for longer periods of time. Thanks to science, we know that most foods are subject to attack by molds, bacteria, and yeast. The degree of susceptibility depends on the food's moisture content. Cereal grains, nuts, and seeds, for example, contain so little moisture that they don't allow microbes to grow. They don't need preservatives. We also know that enzymes, unless stopped, continue to ripen fruit after it is picked and this spoils foods.

So now we bring out the fact that if we freeze these microorganisms and enzymes they become dormant. This slows their growth and activity, thus keeping the food fresh. This led to freeze-drying as a better way to preserve foods.

Increased knowledge of food chemistry and microbiology has given us better ways to preserve foods. Canning plants have laboratories where trained people test samples of the food at every stage. This new knowledge has enabled the food industry to respond to the need for more convenient ways to preserve foods by using chemical preservatives. Sodium benzoate and other benzoates were known to retard the growth and activity of bacteria and enzymes and would keep food fresh. Benzoates are most effective in sour or acid foods which is why they are put in most soft drinks.

To keep fatty foods from going rancid, chemists have developed antioxidants. There are two types of chemical reactions that take place when fats go rancid, oxidative and hydrolytic. In hydrolytic reactions, enzymes speed the reactions of the fat with other materials in the food to produce foul-smelling fatty acids. In the oxidative reaction, oxygen joins the double bonds of some of the unsaturated fats to form such compounds as aldehydes and ketones.

Foods containing large amounts of carbohydrates tend to change color when oxidized. Lemon, lime, or pineapple juice prevents rancidity in sliced fruits. This is because of the natural antioxidant, ascorbic acid or vitamin C found in these juices. The chemicals used most widely to control rancidity and browning are butylated hydroxyanisole (BHA) and butylated hydroxytoluene (BHT). Citric acid and phosphoric acid are used often also. Mold inhibitors such as the propionates and sorbic acid keep molds from growing.

Now science has introduced us to a new process, the irradiation of food. The need for this type of preservation arose when the U.S. Navy wanted to reduce refrigeration facilities aboard ships. Low doses of radiation don't kill all of the bacteria in the food, but their levels are lowered so much that the food won't spoil for a long time. Radiation is a good method of food preservation because it doesn't change the taste, odor, or cooking characteristics of the food. There have been no cases reported of food poisoning from irradiated foods.

In this paper I have tried to show how the methods of preserving food have been advanced with the help of science, particularly chemistry. From salting and drying to Louis Pasteur, to sodium benzoate, to atomic radiation, chemistry has not only kept up with the world's problems but has solved them with new methods and ideas.

An urgent question today is how to make more food available for the ever increasing world population. Some of the solutions being proposed include important roles for chemists.

One solution features the apparent inefficiency of feeding grain crops to animals and then eating their meat rather than using the grain directly for human food. To some, a meat-centered diet is an inefficient use of food resources. These people emphasize, for example, that beef cattle yield as meat on the table only one pound of protein for every 20 pounds of plant protein they are fed.

Satisfactory diets must provide adequate quantities of all the building blocks essential to good health. No single food item does this. Eggs come closest to meeting all of our protein requirements. One goal is to find the blend or mixture of foods that will provide all needed protein, will taste good, will be easy to prepare and serve, and will meet the other requirements of successful food items. Textured vegetable protein foods are one effort in this direction. To produce this food item, plant foods rich in protein, soybeans for example, are taken apart to provide the essential protein building blocks. These materials are then mixed with vegetable oils, flavors, and binders such as egg albumen to give the mixture the proper consistency. Sometimes vitamins and minerals are added.

While this mixture is being prepared, various kinds of foods can be used to provide exactly the kinds and quantities of each of the essential protein building blocks that are called for by a "perfect" protein food. If soybeans are low in one amino acid, for example, another food that is rich in this amino acid is included. Wheat, oats, eggs, yeast, or milk, for example, can be called on to provide amino acids that may be in short supply in soybeans. One of the chief contributions of chemistry to this approach has been its ability to determine the kinds and quantities of each compound or element present in a food. Similarly, modern chemistry can describe quite accurately the effects that are produced by an excess or a deficiency of many of the individual building blocks in food.

After the proper mixture of the required ingredients has been prepared, the mixture is cooked to a thick syrup. To convert this syrup to a satisfactory food item is the next problem. To solve this problem, a process was borrowed from the textile fiber industry. The protein mixture is forced through a spinnerette which is immersed in a water solution containing phosphoric acid and sodium chloride. The spinnerette is similar to those used for rayon and may contain as many as 15,000 holes a few thousandths of an inch in diameter. As the protein mixture fibers emerge, they are stretched to improve their tensile strength and washed with plain water. After the excess water is squeezed out, the rope of fibers or tow is cut into short lengths.

A more complete story about the production of spun-fiber vegetable protein products has been told by Dr. Daniel Rosenfield in the June 1974 issue of *Chemtech*. The spun vegetable protein fiber has characteristics quite similar to those of animal proteins. Thus, the fiber can be combined with other kinds of foods in ways that not only allow these foods to contain the required nutritional elements but also to have proper textures.

People may choose between foods which contain either animal or plant produced proteins. Now, however, they can also choose the kinds

Analytical instruments in a hospital serum cholesterol laboratory provide immediate data on how foods affect our health.

Achievement

Food has been made easier to store, distribute, prepare. We understand much about nutritional needs and can supply diets to meet them.

Choice

Poorly balanced diets, malnutrition, and even famine persist. We face hard personal and public decisions about the use of effective procedures for producing and distributing food.

of food they prefer without being constricted by shortages or cost. There is no doubt that more protein foods would be available for people if farmers' products, especially their cereal grains, went into "people food" rather than "animal food." It is equally true that many people prefer meat-centered diets. Perhaps the future of our society and the health and well-being of the world's people will demand that we must make certain choices rather than to be free to make whatever choices we prefer. If so, it is good to know that men and women are working on narrowing the taste, texture, and cost gap between protein foods derived from plants rather than from animals.

Choices and decisions— looking back and looking ahead

It is difficult to say whether past choices and decisions have been wise without the benefit of hindsight. The stories of two products of the chemical industry illustrate this point of view.

The first story begins in Michael Sveda's laboratory where he was look-ing into the fever-reducing properties of certain substances. When he picked up a cigarette that had been set beside some beakers, he noticed that it had a very sweet taste. Out of curiosity, he tasted the contents of the beakers and found the source of the sweet taste to be hexylsulfamic acid. Additional research proved that a group of compounds, the cycla-mates, were 30 times sweeter than sugar but were not metabolized by the body to release energy. This was good news to the diet food industry whose main sugar substitute at the time was saccharin, which to some people leaves a bitter aftertaste.

The cyclamates were used as sugar substitutes chiefly for diabetics and those who couldn't use sugar in the 1940s and 1950s. In the 1960s, how-ever, with millions of people in the United States becoming weight-conscious, a large market for cyclamates appeared in soft drinks and other beverages and foods.

Between 1963 and 1967 U.S. cyclamate production rose from 5 to 15 million pounds. A 1969 Food and Drug Administration fact sheet esti-mated that 75 percent of the people in the United States were consuming the non-nutritive sweetener. At this time, based on a number of investi-gations which indicated that the cyclamates carried no hazard to health, they were on the Food and Drug Administration's list of substances that were "Generally Recognized As Safe."

In 1966 two Japanese scientists determined that cyclamates were metabolized in the body to produce hexylamine, a substance that could cause dermatitis or convulsions when taken internally. Additional investi-gations involving feeding foods containing cyclamates to rats revealed other disorders that seemed to be caused by the cyclamates.

At this point the Food and Drug Administration called on the National Academy of Sciences-National Research Council for advice. In time, the NAS-NRC responded with the statement, "Up to 5 grams a day could be taken of cyclamates without any harm occurring to an adult" but that unrestricted use of cyclamates might be harmful. On this basis, the FDA proposed that labels on foods and beverages containing cyclamates should carry appropriate warning statements.

In 1969 Jaqueline Verrett and Marvin Legator reported that 600 chicks with deformities hatched from 4,000 eggs which they had injected with cyclamates. These deformities were proved to result from chromosome damage caused, apparently, by the cyclamates. Other investigations late in 1969 reported that heavy and continued doses of cyclamates caused bladder cancers in rats. As interpreted by Michael Sveda, however, the "heavy and continued doses" were equivalent to drinking a bottle of cyclamate sweetened soft drink each minute for 8 hours.

On September 11, 1970 the FDA ordered all cyclamate-containing prod-ucts to be banned although existing stocks could be sold. Almost over-night a billion dollar industry was put out of business. With this decision, cyclamates ceased to be a topic of widespread public interest and concern. This did not mean, however, that the investigation of their long-term use was stopped. In the minds of some investigators the decision to ban the use of cyclamates should be reconsidered. Thus in March 1975 the Food

In this medicinal research laboratory for the preparation of new compounds that might potentially be useful in medicinal agents, chemists work with complex structures, including enzymes.

Achievement
Medicines with which physicians can prevent or alleviate disease, relieve crippling pain and anxiety.

Choice
Hard personal decisions regarding the use and abuse of a variety of drugs; hard public decisions on the testing and availability of pharmaceutical chemicals.

114

and Drug Administration asked the National Cancer Institute to review the total question of cyclamates and the origins of cancer.

The story of a second group of products began in the early 1930s. Charles F. Kettering and a group of business-minded scientists realized that household refrigerators would never be possible until someone put together a refrigerant that was safer than the ammonia used in ice-making machines. Kettering's team knew that the required compound had to be nontoxic, nonflammable, noncorrosive, and have a boiling point not far below the freezing temperature of water. By turning to the classified array of facts on the traditional periodic chart of the elements, their attention was drawn to fluorine. Other information and fortunate circumstances led to the idea that dichloromonofluoromethane would have the desired properties. It did, and the problem of a household refrigerant was solved.

Dichloromonofluoromethane is only one of a group of compounds known as the freons. Other freons can be put together by changing the numbers of chlorine and fluorine atoms that have replaced the hydrogen atoms in methane or other hydrocarbon molecules. Since each new kind of molecule has its unique properties, however, each freon finds unique uses. Fluorocarbon-12, another name for the freon that is made by replacing the four hydrogen atoms in the methane molecule with two atoms each of chlorine and fluorine, is particularly useful as a refrigerant. It is used widely in today's refrigerators, freezers, and air conditioners.

After World War II fluorocarbon-11 became a widely used propellant in spray cans and was used as a blowing agent in making foam rubber and plastics. About 4 pounds of fluorocarbons are produced each year per person in the United States. Some 4000 people hold jobs directly related to fluorocarbon production and distribution. The great majority of refrigeration systems, household as well as automotive and commercial, rely on fluorocarbons. More than half of the nearly three billion spray cans sold each year use fluorocarbons as the propellant.

In 1974 the fluorocarbons became caught up in theoretical interactions with the layer of ozone that exists in the upper atmosphere 8 to 30 miles above the earth. Our environment on earth is influenced by the absorption of ultraviolet solar radiation by the ozone layer. Supposedly, all life on earth has developed in and become adapted to an environment influenced by the absorption of ultraviolet radiation by the ozone layer On this basis, significant changes in this process could pose threats to the health and well-being of all living systems.

Furthermore, it is generally believed that the absorption of ultraviolet solar radiation is the chief source of heat in the stratosphere, and the climate of the stratosphere is linked closely with our climate here on Earth. Because the ozone in the stratosphere absorbs ultraviolet radiation and converts it to heat, the temperature does not continue to drop uniformly with increasing distance above the earth's surface. Temperatures as low as minus 60°F exist at the lower boundary and as high as plus 30°F at the upper boundary of the stratosphere.

This nonuniform or temperature "inversion" is believed to serve as a "lid" which slows the vertical mixing of water vapor and other gases through the stratosphere. This means that water vapor cannot rise indefinitely and escape from the earth's surface but is forced to condense or freeze and fall back to earth as rain or snow. Those substances which rise above our atmosphere are caught up in the turbulence, storms and rainfall and, for the most part, are returned to the earth and its atmosphere

In June 1974 Drs. Mario J. Molina and F. S. Rowland published the theory that fluorocarbon molecules which have escaped into the stratosphere convert ozone to oxygen. They argued that the bonds which hold fluorocarbon molecules together are uniquely susceptible to attack by the light waves found in ultraviolet solar radiation. Furthermore, when fluorocarbon molecules come apart, the unattached chlorine atoms bump into ozone molecules and convert ozone to oxygen. Although ozone molecules differ from oxygen molecules only by having an additional oxygen atom, this is sufficient difference to affect the absorption of ultraviolet radiation.

As soon as it was announced, the ozone depletion theory captured public interest. Scientists realized that the theory held promise of a better understanding of our total environment. Many men and women set out to obtain data that would enable them to measure precisely the events and conditions that were involved in the theory.

At least 15 governmental agencies launched new research or added the ozone depletion theory to ongoing research involving the upper atmosphere. Proposed legislation was introduced in Congress and at least 10 state legislatures.

Those who produced and distributed fluorocarbons stepped up their research programs involving all aspects of the effects of fluorocarbons on the environment. In a program administered by the Manufacturing Chemists Association and carried out by scientists at universities in the United States, Canada, and Great Britain, the combined facilities of many agencies are being brought together. These facilities are being used to provide precise measurements, to test tentative hypotheses and models, and in general pin down the facts needed to prove or disprove the ozone depletion theory.

1776, 1876, 1976, on and on, and on and on

The theme of hopes and fears provides an appropriate note on which to close these glimpes of what chemistry is, what chemists do, and what the results have been. Chemists, as people, have all of the interests and concerns, hopes and ambitions that other people have. Everybody shares the same world.

To help people solve their material problems is the continuing responsibility of chemistry. Any material need can attract the attention of chemists and the chemical industry. Life today, especially in the more affluent nations, is characterized by abundance of materials. Drugstore, supermarket, hardware, and all kinds of shopping center shelves are loaded with items which promise to make life ever more beautiful, comfortable, healthy, or easier for everyone who can afford their prices. Each new item arrives amidst an advertising campaign and then settles down to a niche in the total economy.

Sometimes after a product has settled down in the economy it is brought back into the spotlight because its use is accompanied by unforeseen side effects. DDT was an example. But insect pests must be controlled. No gardener takes pride in rows of vegetables with leaves eaten by insect larvae. Nothing is more pitiful than a young lamb whose flesh is being eaten away by maggots flourishing in untreated wounds. No farmer wants to deliver grain to flour mills or breakfast food manufacturers if tiny weevils are hidden in wheat or rice grains. No public health official dares risk a malaria epidemic because mosquitos were not controlled.

In many choices and decisions, chemists must join forces with governmental agencies. Usually, the problems that are involved are more vital than controlling the production and distribution of chewing gum but chemists are expected to contribute their knowledge and take part in decisions whenever questions arise involving private enterprise interacting with governmental control.

Few people today know the horror of wide-sweeping diseases which no one knows how to avoid, prevent, or cure. Lurking in the background, however, is the awful uneasiness lest some unforeseen factor arise and we know once again the tragedy of mass sickness and death. Typhoid fever, polio, and streptococcic infections are almost nothing more than memories of past years. But to maintain our health and to help find new materials to cure or prevent all diseases continue to challenge the chemical profession and industry.

It is a brute fact that the number of people the world must feed is increasing all of the time whereas the amount of land for growing their food is finite. Hungry people create serious problems, and not only from humane points of view. Arguments for or against psychological and economic reasons for developing diets for our increasing population cannot relieve the chemical profession and industry from the responsibility to look for ways to produce food more efficiently.

The threat of an impending energy shortage holds great concern on the part of chemists. Energy is at the very center of everything they do. Furthermore, oil, gas, and coal provide enormous quantities of the raw

116

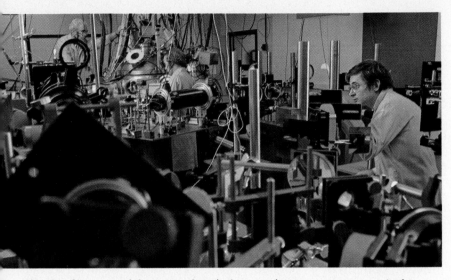

The development of thermonuclear fusion as a future energy source is the task of this Laser Fusion Feasibility Project laboratory at the University of Rochester. The 150-foot-long room is filled with millions of dollars worth of lasers, lenses and supportive electronic equipment.

materials from which the chemical industry makes the things we need. In some cases, it is easier to find a substitute source of energy for fossil fuels than it is to find other sources of the building blocks now obtained from oil, gas, or coal.

Chemists also realize that people want to live ever more abundantly, to travel farther and faster, and to see and do more and more. But the earth's energy resources are finite and the fossil fuels are nonrenewable. Serious problems must be solved before alternative sources of energy can provide us with the quantities of energy we have become accustomed to.

Equally complex problems must be solved before nuclear energy can become generally available. Much has been learned about releasing and controlling the energy that is produced by atomic fission and fusion. Difficult choices and decisions remain, however, before fusion reactors can provide us with the energy we need day after day, year after year, century after century.

Chemists are being called upon increasingly to help solve communication problems. Each of our senses involves energy in some form interacting with highly specialized organs, tissues, and cells. Within these cells are equally highly specialized molecules—specialized in responding to stimuli and transmitting signals. Communication systems based on electrons and photons are even more "fine grained" than are these molecules. This means that these systems hold promise of providing truly faithful reproductions of things being seen, sounds being heard, smells being smelled, tastes being tasted, and textures being felt.

There is another set of choices and decisions which determines whether people can or cannot communicate effectively. There are people whose sensory systems work well but they lack the related experiences and sensations they need to make meaningful experiences out of newly arriving messages. Chemistry teaching provides many opportunities to exercise and to study communication phenomena and these opportunities enable the chemical profession and industry to assume increasingly significant responsibilities in education.

Chemists have always shared our hopes for a better world as well as the haunting fears of the consequences of problems not solved, needs not fulfilled, actions that must be taken without complete knowledge of their ultimate outcomes.

What of the future? It would be good to point to new and better materials, to new and ever more revealing insight into the mysteries of ourselves and our environment. These will come. We need no gazing into crystal balls. As long as people are free to pursue whatever their curiosity invites, as long as people are free to create new and recycle existing knowledge, the achievements of the past will go on, and on and on.

ACKNOWLEDGMENTS

The author acknowledges the inspiration and feedback provided by his students at the Charles W. Woodward High School in Montgomery County, Maryland. He is especially grateful to Lynn Ludmer, William Parler, and William Savarino whose papers are included in the book, to Richard Heyman and Douglas Ziegler who are quoted in the Preface, and to Cathy Bennett, one of his laboratory assistants who helped develop and criticize the manuscript.

Among the people who reviewed portions of the manuscript are Dr. Theodor Benfey, Guilford College and Editor of *Chemistry;* Dr. William F. Kieffer, The College of Wooster; Gail Nussbaum, Northwestern High School, Hyattsville, Maryland; Dr. Joseph S. Schmuckler, Temple University; and Dr. Robert E. Varnerin, Manufacturing Chemists Association.

Among the many people and organizations who provided essential information are James Abrams, Director, Public Relations, Monsanto Company; Dr. Glenn R. Brown, Vice President, Research and Engineering, Sohio; Stephen M. Downey, FMC Corporation; Dr. J. D. D'Ianni, Goodyear Tire and Rubber Company; George L. Gaines, Jr., General Electric Research and Development Center; Dr. Elbert E. Gruber, Vice President and Director of Research and Development Division of General Tire and Rubber Company; Dr. Henry B. Hass, Chemical Consultant; T. L. Heying, Director of Research, Olin Corporation; Dr. P. C. Kearney, United States Department of Agriculture; L. F. McBurney, Director, Research Center, Hercules, Incorporated; Dennis D. Mog, Product Program Manager, Corning Glass Works; Dr. A. H. Reidies, Director of Research, Carus Chemical Company; Dr. John R. Thirtle, Eastman Kodak Company; and Harold Wittcoff, General Mills Chemicals.

The author also acknowledges the aid provided by American Chemical Society staff members Joan Comstock, Mary Rakow and Charlotte Sayre for editing and proofing.

Finally, the author benefitted from the deliberations of the American Chemical Society Centennial Exhibit Committee.

THE AUTHOR

John H. Woodburn has degrees from Marietta College, Ohio State University, and Michigan State University. He began science teaching in 1936. Following service in World War II, he began doctoral work in science education at Michigan State University. He had the unique advantage of a doctoral committee composed of members from the Education Department and from academic departments including Dr. L. L. Quill, Chairman of the Chemistry Department.

Dr. Woodburn has held teaching positions at Michigan State, Illinois State University, and the Johns Hopkins University. He has also taught evening or summer courses for Oregon College of Education, Syracuse University, the University of Virginia, George Washington University, and the University of Maryland. In 1953-56, he served as the Assistant Executive Secretary of the National Science Teachers Association and in 1957, as Science Consultant with the U.S. Office of Education.

In 1960, he returned to chemistry teaching and began writing science-related books for young people. Among his titles are *Radioisotopes* (Lippincott, 1962), *Excursions Into Chemistry* (Lippincott, 1965), *Opportunities in the Chemical Sciences* (Vocational Guidance Manuals, 1971), and *The Whole Earth Energy Crisis* (Putnams, 1973). He also co-authored *Teaching the Pursuit of Science* (Macmillan, 1964) and *Demonstrations and Activities for High School Chemistry* (Parker, 1971).

Dr. Woodburn received Science Teacher Achievement Recognition Awards in 1969, 1971 and 1975, the National Science Teachers Association Citation for Distinguished Service to Science Education in 1971, and the Manufacturing Chemists Association National Award for Excellence in Chemistry Teaching in 1975. He has been at the Charles W. Woodward High School in Montgomery County since 1966.

PICTURE CREDITS:
1—B. Riley McClelland. 2—(top, bottom) National Geographic Society, (middle) United States Department of Agriculture (USDA). 17—(top) St. Luke's Hospital Center, (bottom) and page 26—Dr. Arthur E. Girard, Pfizer, Inc. 37—USDA. 38—National Geographic Society. 44—USDA. 46—USDA. 55—National Aeronautics and Space Administration (NASA). 56—National Geographic Society. 66—Eric Poggenpohl. 69—Du Pont. 74—NASA. 83—National Oceanic & Atmospheric Administration. 89—AGA Corporation. 90—(top right, bottom left and right) Eastman Kodak Company. 91—NASA. 94—(left top) NASA, (right top and bottom) Eastman Kodak Company.

Index

Propylene, 15
Proteins, 20, 23, 47, 61, 64, 113
Prothrombin, 31
Protons, 8
Proust, Joseph, 81
Pyrethrum, 52
Pyrolusite, 63

Q

Qualitative analysis, 63
Quantitative analysis, 64

R

Radioactive wastes, 106
Radioactivity, 10
Radium, 11
Rancidity, 112
Raw materials, 4
Rayon, 68
Reade, J. B., 91
Recycling, 111
Reforming hydrocarbons, 78
Residue build-up, 54
Respiration, 44, 47
Rickets, 31
Rinne, Robert W., 47
Robbins, F. C., 28
Rosenfield, Daniel, 113
Rosing, Boris, 95
Rotenone, 52
Rotheim, Erik, 61
Rowland, F. S., 115
Rubber, 5, 66, 68, 106

S

Sabin, A. B., 28
Saccharin, 114
Salk, Jonas, 29
Salmonella, 108
Salt, 7, 112, 113
Sandstead, Harold H., 51
Sapodilla tree, 106
Schizophrenia, 34
Schrader, Gerhard, 53
Schweitzer, 68
Scouring powder, 70
Screen printing, 99
Scurvy, 31
Selective pesticides, 104
Selenium, 50, 97
Semiconductors, 96
Sensitivity, 26
Sequestering, 70
Serotonin, 36
Sewage, 48, 62
Sex attractants, 105
Sheehan, John C., 26
Shellac, 61
Side-effects, 33
Silicon dioxide, 57
Silicone oils, 61
Silicone rubber, 68
Silver, 58, 91, 96
Silver chloride, 58, 84
Silver iodide, 84
Silver nitrate, 91
Silver oxide, 84
Slife, Fred W., 47
Snyder, Solomon H., 36
Soaps, 70
Sodium, 49, 80
Sodium aluminum acid phosphates, 71
Sodium benzoate, 112

Sodium carbonate, 57, 60
Sodium chloride, 8
Sodium hydroxide, 68, 98
Sodium pyrophosphate, 70
Sodium thiosulfate, 91
Soil, 48
Solanine, 40
Solar energy, 41, 46, 83, 117
Solids, 13
Solutions, 13
Solvay, Ernest, 8
Soybeans, 46, 113
Spandex, 69
Spirochetes, 24
Spodumene, 59
Spontaneous generation, 112
Spray cans, 61, 115
Spruce gum, 106
Sprue, 31
Spun-fiber protein, 113
Starch, 43, 64, 70
Static electricity, 99
Stein, L., 34
Stookey, S. D., 58
Storage batteries, 83
Streptococcal bacteria, 24
Styrene-butadiene, 68
Substitute natural gas, 79, 83
Succinylcholine chloride, 33
Suckow, 99
Sugar, 6, 21, 64, 82, 114
Sulfanilamide, 24
Sulfa drugs, 24, 28, 36
Sulfur, 66, 79
Sulfur dioxide, 80
Sulfuric acid, 50, 59, 68, 70, 84
Superphosphate, 50
Superstition, 24
Survival Technology, 58
Sveda, Michael, 114
Symbols, 5
Synthetic rubber, 59, 66, 106
Syphilis, 24
Szent-Gyorgi, Albert, 16

T

Talbot, William Henry Fox, 91
Temperature inversion, 115
Tempered glass, 58
Terylene, 69
Tetraethyl pyrophosphate, 53
Textured vegetable protein, 113
Thermal copying, 99
Thorium, 10
Thrombin, 31
Thyroid hormone, 34
Titanium dioxide, 61
Titanium tetrachloride, 66
Tortoise shell, 16
Toxic substances, 40
Trace elements, 19, 50
Tranquilizers, 36, 59
Transistors, 96
Transuranium elements, 13
Triacetates, 69
Trichloromonofluoromethane, 61
Triethylaluminum, 66
Trisodium phosphate, 70
Tritium, 86
Trona, 7
Trypanosomes, 24
Trypsin, 36
Turpentine, 100
Typhoid, 108
Typhus, 52, 103
Tyrosinase, 23

U

Ultramarine, 60
Ultraviolet radiation, 115
Underground gasification, 79
Unsaturated hydrocarbons, 76
Uranium, 10, 86, 106
Uranium hexafluoride, 88
Urine, 22, 64, 69
Urobilinogen, 64

V

Vaccine, 28, 29
Valence, 76
Valium, 34
Vanadium, 50
Van Campen, Darrell, 50
Varnish, 64, 86
Verrett, Jaqueline, 114
Videotape, 98
Vinegar, 6
Vinyl acetate, 61
Vinylpyrrolidone, 61
Virus, 25, 27
Viscose, 68
Vitamin A, 59
Vitamin C, 112
Vitamin K, 31, 36
Vitamins, 31, 45
Vogel, Herman C., 93
Von Mering, J., 21
Von Sachs, Julius, 51
Vulcanization, 66

W

Warfarin, 31
Warren, J. C., 32
Washing soda, 7, 70
Wastes, 106, 107
Water, 70, 72
Water pollution, 63, 70
Watson, William I., 79
Weller, T. H., 28
Whale oil, 76
Wharton, F. D., Jr., 109
Whipple, George, 31
Williams, G., 67
Wintergreen, 6
Wise, C. D., 34
Wöhler, Frederick, 11
Woodward, John, 51
Woodwell, George M., 103
Wurster, Charles F., Jr., 103

X

Xerography, 99
Xerophthalmia, 31

Y

Young, Thomas, 92

Z

Zeidler, 52
Zelitch, Israel, 47
Ziegler, Karl, 65
Zinc, 50, 83
Zinc hydroxide, 84
Zuelzer, G., 21
Zworykin, Vladimir, 95

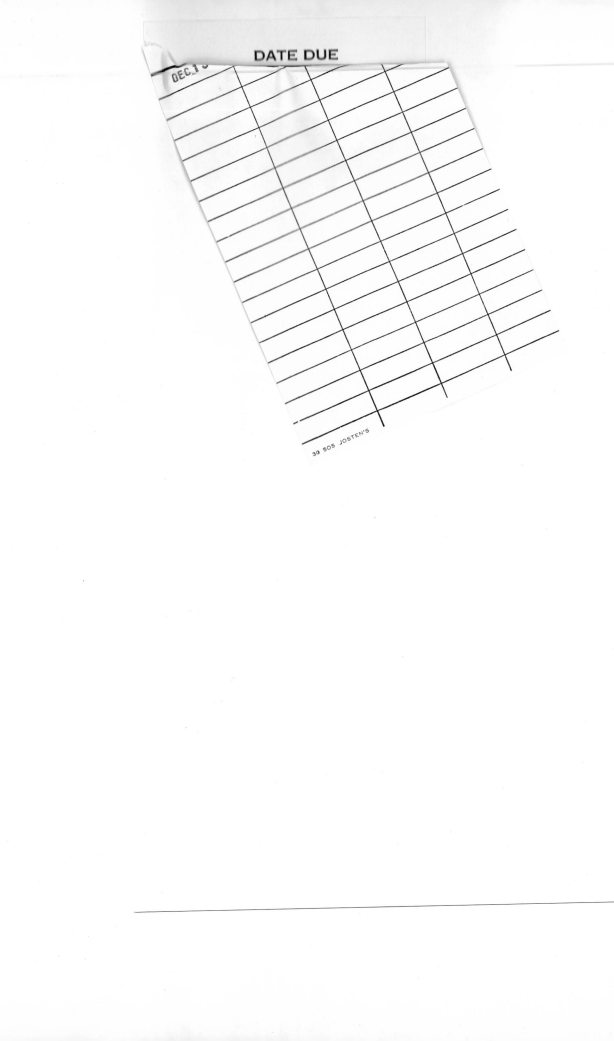

DATE DUE

DEC 1

30 505 JOSTEN'S